# ELECTROINDUCED DRIFT OF NEUTRAL CHARGE CLUSTERS IN SALT SOLUTIONS

# ELECTROINDUCED DRIFT OF NEUTRAL CHARGE CLUSTERS IN SALT SOLUTIONS

Li Hunda

M. A. Kazaryan

I. V. Shamanin

CRC Press
Taylor & Francis Group
Boca Raton  London  New York

CISP

CRC Press is an imprint of the
Taylor & Francis Group, an **informa** business

Translated from Russian by V.E. Riecansky

# Contents

# Foreword

The formation and evolution of unsteady electric fields, as well as their interaction with matter, are part of the general process of redistribution of electromagnetic energy in space which is initiated by various reasons. The action of an external electromagnetic field on a solution (by definition) increases the energy stored in the volume of the solution.

The energy reserve of a salt solution in a polar dielectric liquid in a state of equilibrium is distributed over three energy 'reservoirs' corresponding to translational, rotational and real degrees of freedom (we consider temperatures such that electronic degrees of freedom are not excited). These reservoirs communicate, and energy can flow relatively freely from one reservoir in another (energy exchange between different degrees of freedom occurs in collisions of solvated ions).

Suppose now that we have changed (for example, increased) the energy supply in one of the tanks. It is clear that excess energy will immediately begin to overflow into neighboring ones. So there is an idea of nonequilibrium (irreversible) processes, which are due to energy exchange in collisions. Irreversible processes leading to the establishment of a statistical equilibrium in the system are referred to as relaxation processes.

Nonequilibrium in a solution occurs if the solution is exposed to external influences. Imbalance occurs in a weak electric field. It is significant that small perturbations of the equilibrium distribution function in principle can determine not the corrections to the effects, but the effects themselves. The relaxation time of the energy of an external electric field at the vibrational degrees of freedom of solvated ions are incommensurably longer than the relaxation time on the vibrational degrees of freedom of the molecules that form the solution. Studies have shown that it reaches milliseconds. The effect of electroinduced selective drift of solvated ions in salt solutions in polar dielectric liquids proceeds with characteristic times and spatial scales commensurate with the corresponding scales of relaxation processes in the solvation shell. This allows counting on the discovery of new significant nonequilibrium effects. If we take into account also the unresolved problems of electromagnetic self-energy, then we can conclude that experimental and theoretical studies of processes accompanying the interaction of unsteady electric, magnetic fields and electromagnetic waves with solutions (taking into account the existence and characteristics of the internal structure of solutions) will still bring new and even unexpected results for a long time.

In interpreting the experimental results, one can rely on the Debye–Hückel approximation, which is widely used in solution theory, in which many useful and 'working' results are obtained, or proceed from the general principles of the theory of electricity and magnetism. These provisions, generally speaking, do not require the separation of charges into positive and negative, with the same varieties charges are repelled, and different one attracted. Rather, they require that the positive and negative charges be 'mixed equally', and the balance between them is so perfect that next to this 'mixture' no action of forces is felt. Any the electromagnetic effect on such a mixture causes deviations from perfection inside the mixture, and these deviations are such that they compensate for the external action, striving for the former perfect in a system that will include both the object of action and self-action.

The book provides information obtained in studies of the processes accompanying the effects of periodic electric and magnetic fields on salt solutions in polar dielectric liquids. The authors tried to focus the explanation of phenomena from a physical point of view, omitting the details of theoretical constructions and mathematical calculations. This is done so that the book is accessible to a wide audience, including students of an experimental profile, and in order to help the reader navigate the multilateral topic that is involved in the study of processes occurring in liquid environments under the external influence of an electromagnetic nature. The authors hope that it will contribute to the development of nonequilibrium molecular physics as a science of physical and chemical processes.

*Lee Hunda, M.A. Khazaryan, I.V. Shamanin*

# Introduction

The action of electric and magnetic fields, as well as electromagnetic waves on a substance is the subject to fundamental and applied research throughout the history of mankind. In any case, at all times of the existence of man, questions arose reasonably caused by the consequences of the action of the same solar radiation on human reality. Such studies will remain fundamental until it will not become clear inwhat cases it can be considered that the current on matter, the electromagnetic field will slightly change the properties of the substance. The assumption that the external field does not change the properties substances means that the microfields acting inside the substances are considered very large compared to the external field and the influence of this field per substance is not large. In this case, the possibility remains of a phenomenological description of a substance under the action of external fields using experimentally determined nonlinear susceptibilities [1]. However, there are effects that indicate the fact that the simultaneous action of two fields on a substance leads to a result that does not match the result of an independent action each of these fields per substance. To explain this, a nonlinear electrodynamics apparatus was developed. It should be noted that the question outlined above is still open.

The solution is a mixture of at least two substances. It should be noted that the action of the same field individually on these substances may not cause any changes. their properties, and the action of the same field on a mixture of these substances leads to irreversible changes. To describe this, the usual apparatus of physics is used, for example, which is the basis of electrochemistry. Note that the structure of the solution as such does not play a special role in the description of the observed changes. The theory of electricity is inviolable [2], and its simple transposition to electrical phenomena in solutions seems to be quite natural,

So, if we study the effect of external electrical, magnetic and electromagnetic fields on solutions, the following questions will inevitably arise:

- first – which fields and what intensity change the properties of the solution;
- the second – how the field acts on a mixture of substances, or rather, how
- the field acts on each of the substances that are mixed and interact with each other;
- the third – what is considered to be a unit of solution structure, if
- its internal structure does not affect the external field, but does it affect the solution as a whole?

Finally, the final, fourth question – what to expect from the action of an external field on the solution?

Attempts, including successful ones, to answer these questions individually or to several of these questions at the same time are the subject of many studies in physics, chemistry, and biology.

These questions arose in the interpretation of the results of experiments conducted by the authors and which were not originally aimed at finding the answer to them.

In experiments whose purpose was to study the effects of external periodic electric fields with a frequency of up to 10 kHz to normal (pH = 7) salt solutions, an interesting effect. It consisted of the following.

The field was created by two flat metal plates and an dielectric vessel filled with an aqueous salt solution was placed between the two. There was no electrical contact between the plates and the solution. A periodic electrical signal was generated by a voltage generator. At certain intervals, samples were taken of the solution to determine its acid-base properties. They remained virtually unchanged.

In one experiment, it was noted that a solution located closer to one of the plates acquired alkaline properties. After the generator was turned off the next day, the solution had normal properties. Turning on the generator and the effect of the field on the solution during a few hours again caused the appearance of alkaline properties in the sample taken closer to the plate. Previously this solution behaviour was not noted.

It took quite a while to find out the reasons, but one of the reasons was simple: a generator malfunction led to the fact that starting from a certain point in time, the generator formed a distorted periodic signal: one of the half-periods of a sinusoidal electrical signal had a smaller amplitude.

This accident led to the fact that the action of the external period of the electric field, a feature of which was the inequality of the voltage amplitudes in half-periods, for solutions of salts in water, and then in other dielectric fluids it became the subject of research by the authors. The experiments gave rise to new questions and searches for answers to them led to new theoretical results, which, in turn, led to new experiments and the results of theoretical and experimental parts of these studies have been preset in this book.

Most applications of various physical phenomena and processes in science, engineering and technology are based on the fact that with a chain of successive events takes place in any initiating action, leadin eventually to the desired result. In such a chain of events there is necessarily a link in which the substance is involved as such in a particular physical state and aggregate form. A substance may simply be an intermediary. In this case, having fulfilled its function, it will remain the same substance. A simple example of this is the substance of the conductor through which a current is transmitted that does not cause its significant heating. In another case, the substance, performing its function, undergoes qualitative changes. The same conductor material through which a significant current flows can pass into a plasma state in a few microseconds. The so-called *electrical explosion of conductors* occurs. The substance can repeatedly perform the function of a conductor of directional electron motion in a limited volume in the first case, and in the second case – only once. In terms of techniques and technology it is convenient that the substance performs its function many times, while maintaining its properties before the next action.

Studies performed by the authors suggest that the solvent substance acts as an intermediary. Solvent molecules allow the external field to act in an unusual way on structural units (atoms, molecules) of the dissolved substances [3]. In chemistry, it is widely used, only the external action is caused by not electric, magnetic or electromagnetic field, and other factors. Perhaps those patterns that are found by the authors and presented in the book will be useful when creating new chemical or physico-chemical technologies. In any case, one of the possible applications in rare chemistry, scattered and radioactive elements are reasonably well conditioned and are also described in the book.

# 1

# Periodic action of electric and magnetic fields on the electrically isolated salt solution in polar dielectric liquids

The discovery of the phenomenon of electroinduced selective drift of cationic aquacomplexes in aqueous solutions of salts caused a lot of questions, one of which was the question of the structure of the salt solution in the polar dielectric fluid and the structure of the clusters, formed by ions and associated molecules around them f the solvent. Traditional models and approximations in which many interesting results were obtained, did not give the opportunity to get a quantitative assessment correlating with experimentally observed effects. This triggered a search for new models and approximations. Inertial properties of supramolecular structural units – clusters are determined by their size. The sizes of these clusters range from tens of angstroms up to several microns. Thus, an aqueous salt solution is formed by nanostructures, and biological and many technological systems function with the participation of nanoparticles: for example, a conventional system of blood circulation or many of the chemical technological processes. Random or continuous action of aperiodic electric fields, magnetic fields and electromagnetic waves on systems, containing salt solutions, causes either positive or negative effective effects. Very often we encounter them, but do not pay due attention to finding out their causes.

## 1.1. Oscillations of ions in a salt solution under the action of external periodic electric field

The practical use of the detected phenomenon of the selective drift of solvated cations of various metals in solutions of salt mixtures under the action of an external periodic electric field requires a clear qualitative picture of the phenomenon. In particular, to determine the speed of directional electroinduced solvated ion drift, it is necessary to determine the parameters of its oscillatory motion caused by the action of an external field. The solvated ion is in a polar dielectric fluid, and the electric field acts not only on it, but also on the molecules of the solvent. In turn, the intrinsic electric field acts on the solvent molecules located in the vicinity of the ion. The task of determining an equation that describes the ion's vibrations with respect to the solvent molecules surrounding it is not trivial. To determine such an equation, we consider the relationship of electric fields and currents corresponding to some rows in a salt solution in a polar dielectric fluid.

### 1.1.1. Action of an external electric field on free and associated charges in solution

Tension distribution of the electric field $E(r)$ and bulk density distribution charge $\rho(r)$ are related by the Poisson equation [4]

$$\operatorname{div} E\ (\mathbf{r}) = 4\pi\rho\ (\mathbf{r}).$$

If the salt the molecules of which are dissociated is dissolved in a polar dielectric fluid, then the electric field in the volume of the liquid is formed by free charge carriers – ions, forming during dissociation, with the distribution of volumetric charge density $\rho_f$ (r). The intensity distribution of this field is also described by the Poisson equation

$$\operatorname{div} E_f\ (\mathbf{r}) = 4\pi\rho_f(\mathbf{r}).$$

Molecules of a dielectric fluid (solvent) are polarized in the electric field of free charge carriers. In this case, the condition [5] is fulfilled

$$\operatorname{div}\ (E\ (\mathbf{r}) + 4\pi P\ (\mathbf{r})) = 4\pi\rho\ (\mathbf{r}), \tag{1.1}$$

where $\mathbf{P}(\mathbf{r})$ is the distribution of polarization density. Moreover, the condition is fulfilled regardless of how $\mathbf{E}(r)$ and $\mathbf{P}(r)$ are related.

The application of an external electric field to the solution leads to a violation of its local electroneutrality, that is, to the appearance of a nonzero bulk charge density. These charges are connected and arise solely due to heterogeneous polarization of dielectric fluid molecules. Distribution of the volume density of bound charges $\rho_b(r)$ is determined by the relation

$$\rho_b(\mathbf{r}) = -\mathrm{div}\mathbf{P}\,(\mathbf{r}). \tag{1.2}$$

In the general case, the space charge consists of two components - the density of free charges and the density of bound charges, which arise in the presence of an inhomogeneous polarization of the sample $\mathbf{P}(\mathbf{r})$.

For simplicity, we restrict ourselves to considering the one-dimensional case (when the solution is in the form of a plane parallel layer, whose transverse dimensions are much larger than its thickness). Polarization density is connected with the strength of the external electric field $\mathbf{E}_0$, formed by the the plane-parallel electrodes, isolated from the solution, by the ratio

$$4\pi\mathbf{P} = \mathbf{E}_0 - \mathbf{E},$$

where $\mathbf{E}$ is the field strength inside the solution layer located between the electr.

The field strength inside the solution is distributed in a certain way, that is, it depends on the coordinate. Moreover, the averaged value of tension in the volume of solution $\langle E \rangle$ is less than the field between the electrodes $E_0$. The ratio $E_0/E$ can be estimated using equation (1.1) and the relationship between the polarization density and the field strength causing polarization of the dielectric liquids:

$$\mathbf{P} = \frac{3}{4\pi}\left(\frac{\varepsilon-1}{\varepsilon+2}\right),$$

where $\varepsilon$ is the dielectric permittivity of the solvent.

If the electric field is formed by a charge with a bulk density $\rho$, and the polarization is caused by the electric field of this charge, then equation (1.1) can be rewritten in the form

$$\left(1+3\left(\frac{\varepsilon-1}{\varepsilon+2}\right)\right)\operatorname{div}\mathbf{E}=4\pi\rho.$$

This equation shows that at the same charge density $\rho$ the polarization reduces the electric field strength $\left(1+3\left(\frac{\varepsilon-1}{\varepsilon+2}\right)\right)$ times.

For large dielectric permeability ($\varepsilon > 10$) the field strength in the volume of dielectric liquid $\langle E \rangle$ is approximately 4 times less the tension between the electrodes $E_0$. When the dielectric constant is close to 1, the field strength in the volume of the dielectric fluid $\langle E \rangle$ is approximately 2 times less between the electrodes $E_0$.

If we talk about the distribution of field strength, then from the much higher considerations it follows that the tension increases in the volume of the plane layer of the solution from the value of $\langle E \rangle$ to the value of $E_0$ when approaching the electrodes that are isolated from the solution. The macroscopic value of the density of bound charges $\rho_b(\mathbf{r})$ will be nonzero only in these near the electrode areas of the solution, since at a distance from the electrode the field strength will depend weakly on the coordinate and have the value of $\langle E \rangle \approx$ const. Moreover, the polarization density $\mathbf{P}$ also will not noticeably change, and $\operatorname{div}\mathbf{P}(\mathbf{r})$ will be close to zero. Indeed, the macroscopic value of the density of bound charges $\rho_b(\mathbf{r})$ due to equality (1.2), as it moves away from the electrodes, it also tends zero.

Taking into account the above reasoning, we can write the equation relating the electric field strengths in the volume of the solution with the density of free charges in the salt solution in a polar dielectric fluid:

$$\operatorname{div}\left(\mathbf{E}_f(\mathbf{r})+3\left(\frac{\varepsilon-1}{\varepsilon+2}\right)\left(\mathbf{E}_0(\mathbf{r})+\mathbf{E}_f(\mathbf{r})\right)\right)=4\pi\rho_f(\mathbf{r}).$$

For the solution layer situated between the electrodes plane and isolated from the solution, the equation is in the form:

$$\frac{\partial}{\partial x}\left(E_f(x)+3\left(\frac{\varepsilon-1}{\varepsilon+2}\right)\left(E_0(x)+E_f(x)\right)\right)=4\pi\rho_f(x),$$

or, which is the same,

$$\left(1+3\left(\frac{\varepsilon-1}{\varepsilon+2}\right)\right)\frac{\partial}{\partial t}\left(\frac{\partial E_f(x)}{\partial x}\right)+3\left(\frac{\varepsilon-1}{\varepsilon+2}\right)\frac{\partial}{\partial t}\left(\frac{\partial E_0(x)}{\partial x}\right)=4\pi\frac{\partial \rho_f(x)}{\partial t}.$$

## 1.1.2. Equation of oscillations of an ion in relation to solvent molecules

We differentiate the resulting equation in time:

$$\left(1+3\left(\frac{\varepsilon-1}{\varepsilon+2}\right)\right)\frac{\partial}{\partial t}\left(\frac{\partial E_f(x)}{\partial x}\right)+3\left(\frac{\varepsilon-1}{\varepsilon+2}\right)\frac{\partial}{\partial t}\left(\frac{\partial E_0(x)}{\partial x}\right)=4\pi\frac{\partial \rho_f(x)}{\partial t}. \quad (1.3)$$

The right-hand side of expression (1.3), by virtue of continuity, be expressed in terms of current density $j_{ion}$ of free charge carriers

$$4\pi\frac{\partial \rho_j(x)}{\partial t}=4\pi\frac{\partial j_{ion}}{\partial x}.$$

For a single ion in the solution, the spatial intensity distribution of the electric field created by this ion can be set as a function

$$E_f(x)=kq_{ion}\frac{x-x_0(t)}{\left[x-x_0(t)\right]^3},$$

where $k$ is a constant, $q_{ion}$ is the electric charge of an ion, $x_0(t)$ is the coordinate of the ion, which changes in time due to the movement of the ion in an external electric field. Then

$$\frac{\partial}{\partial t}\left(\frac{\partial E_f(x)}{\partial x}\right)=kq_{ion}\frac{\partial}{\partial t}\left(\frac{1}{\left|x-x_0(t)\right|^3}-3\frac{1}{\left(x-x_0(t)\right)^3}\right)$$

and

$$\frac{\partial}{\partial t}\left(\frac{\partial E_f(x)}{\partial x}\right)=-6\frac{kq_{ion}}{\left(x-x_0(t)\right)^4}\frac{\partial x_0(t)}{\partial t}=-6\frac{kq_{ion}}{\left(x-x_0(t)\right)^4}v_{ion}(t),$$

$v_{ion}(t)=\dfrac{\partial x_0(t)}{\partial t}$ is the speed of the directed motion of ions. Bearing

in mind that the ion current density $j_{ion}=nq_{ion}v_{ion}$, where $n$ is ion density, we write equation (1.3) for the coordinate of a single ion $x_i(t)$

in a salt solution in a polar dielectric liquid with a high dielectric constant:

$$-24\frac{kq_{ion}}{(x-x_i(t))^4}\frac{\partial x_{xi}(t)}{\partial t}+3\frac{\partial}{\partial t}\left(\frac{\partial E_0(x)}{\partial x}\right)=4\pi\frac{\partial n}{\partial x}q_{ion}\frac{\partial x_i(t)}{\partial t},$$

or

$$\left(4\pi\frac{\partial n}{\partial x}+24\frac{k}{(x-x_i(t))^4}\right)q_{ion}\frac{\partial x_i(t)}{\partial t}=3\frac{\partial}{\partial t}\left(\frac{\partial E_0(x)}{\partial x}\right).$$

Let us analyze the last equation. The numerical value of the constant $k$ (proportionality coefficient in Coulomb's law) in the system of the physical quantities SI is $k \approx 9 \cdot 10^9$. We choose a fixed point in the solution volume $x = a$, at which the gradient of the field strength is known and its time dependence $\partial E_0(a)/\partial x = f(t)$ In experiments with the excitation of selective drift of the solvation ions [6] $f(t) \approx A (1 + 2 \sin \omega t - \sin 2\omega t)$, the constant is $A \approx 4 \cdot 10^6$ V m$^{-2}$. The amplitude of oscillations of the ion $\Delta(t)$ in the solution under the action of an external electric field $(a - x_i (t))_{max}$ is known not to exceed $10^{-3}$ m, therefore, the first term in the parentheses in the brackets of the left parts of the equation can be neglected. The equation for the oscillations of the ion $\Delta(t)$ is written as:

$$24kq_{ion}\frac{1}{\Delta(t)^4}\frac{\partial}{\partial t}\Delta(t)=6A\omega\left(\cos\omega t-\cos 2\omega t\right),$$

or

$$\frac{1}{\Delta(t)^4}\frac{\partial}{\partial t}\Delta(t)=\frac{A\omega}{4kq_{ion}}\left(\cos\omega t-\cos 2\omega t\right). \tag{1.4}$$

The solution of equation (1.4) has the form [7]

$$\Delta(t)=\left(\frac{3A}{8kq_{ion}}\left(\sin 2\omega t+2\sin\omega t\right)+C\right)^{-1/3}, \tag{1.5}$$

where the constant $C$ is determined from the condition $\Delta(t = 0) = a$.

Oscillation frequency $v = \omega/2\pi$ coincides with the frequency of the external electric field and represents the frequency of the oscillations of the ion relative to the solvate shell. If this frequency coincides with its own frequency of the solvated ion, i.e., the ion–solvate shell

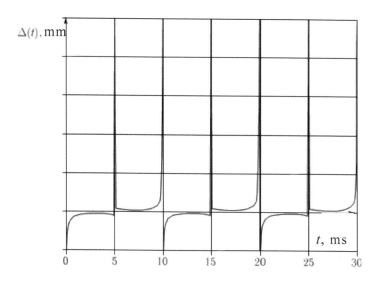

**Fig. 1.1.** Oscillations of a calcium ion inside a sphere formed by molecules of water in the solvate shell, under the action of an external electric field.

system, then we should expect the transition of the vibrational motion into the translational drift of the solvated ion [3]. The intensity of the translational motion will be proportional to the difference in the amplitudes of the deviation of the ion from the initial position during the positive and negative half-periods of the oscillations of the external electric field.

The results of using the obtained equation of oscillations of the central ion located inside the solvation shell are shown in Figs. 1.1 and 1.2.

The obtained equation of ion vibrations in solution for the case of 'asymmetric electric field' (1.5), when the amplitude and durationof the half-periods are different, shows that the ion displacement relative to the initial position for one period has an alternating character. Moreover, the absolute values of the displacements for the 'positive' $\Delta_+$ $(0 \leq T/2)$ and 'negative' $\Delta_-$ $(T/2 < t < T)$ half-periods are also different. The difference between them $l_d = |\Delta_+| - |\Delta_-|$ represents a segment of the ion drift trajectory in the direction of one of their electrodes, which form a field in a flat layer of a solution. A situation similar to the 'two steps forward, one step backward' pattern is realized and, as a result, the ion drifts directionally in one direction. Parameters of the trajectory formed by successive displacements of the ion in opposite directions using the resulting

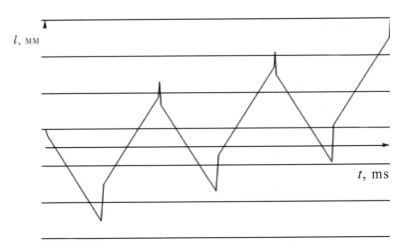

**Fig. 1.2.** The calculated trajectory of calcium ion in an aqueous solution under the action of external electric field.

equation, are described only at a qualitative level. The reason is that in the formulation of the problem, the friction of the solvated ion is not taken into account with the surrounding solvent molecules. This leads to the result that the the amplitudes of displacements of the solvated ion are much lower than the amplitudes of displacements of the ion inside the solvation shell.

The application of the obtained equation is well conditioned for calculating the parameters of ion vibrations inside the solvation shell. The value of the constant $A$ included in the solution is determined to a greater extent not but by the amplitude of the external electric field, but by its distribution in the solvent volume in the interelectrode gap, i.e. by the properties of a solvent as a polar dielectric. The greaterthe gradient of tension in the interelectrode gap, the more significant value is $A$.

### 1.1.3. Consideration of friction with a solvated ion with the molecules surrounding the solvents

The resulting effect caused by the friction of the ion oscillating in the external field with the surrounding molecules can be quantified. To do this, we write the equation of motion for an ion that is affected by an external force $F$, which varies cosine in time, as well as a restoring force and force
resistance (friction):

$$\ddot{x} + 2\lambda\dot{x} + \omega_0^2 x = \frac{F}{m}\cos\gamma t, \qquad (1.6)$$

where $m$ is the mass of the ion; $\gamma$ is the frequency of change of the external electric field creating the force; the second term takes into account the friction force proportional to the velocity; the third term takes into account the return force proportional to the displacement from the equilibrium position. The friction force, acting on a solvated ion, can be determined according to the Stokes law

$$F_1 = 6\pi r\eta\dot{x},$$

where $r$ is the radius of the solvated ion, $\eta$ is the dynamic viscosity solvent. Thus, $\lambda = 3\pi r\eta/m$. The $\omega_0^2$ value is defined as $\omega_0^2 = k/m$, where $k$ is a coefficient similar to the coefficient of rigidity for a spring, if we consider a cluster formed by a central ion and solvent molecules associated around it, as a hollow spherical shell with inner radius $R_1$, outer radius $R_2$ and mass $M$. In this case, the hollow spherical shell can rotate about an axis passing through its geometric centre. The shell is connected to this elastic spiral axis spring by stiffness $k$, which provides a stable position of the shell. In this model, the cluster is considered as a torsion spherical pendulum. The stiffness of the spring is determined by the binding energy of the central ion with solvent molecules located on some distance from it. These molecules form the first radius solvation of $R_1$, and their number is equal to the coordination number of the ion in a given solvent.

The natural frequency of such a pendulum is determined according to the formula

$$v = \sqrt{\frac{k}{J}},$$

where $J$ is the moment of inertia of the pendulum, the value of which is determined according to the formulas

$$J = \frac{2}{5}MR_2^2 \qquad \text{for the case } R_2 \gg R_1;$$

$$J = \frac{2}{3}M\left(\frac{R_1 + R_2}{2}\right)^2 \quad \text{for the case } R2 \sim R_1.$$

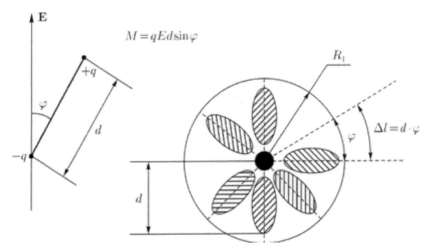

**Fig. 1.3.** The occurrence of a moment of force when deviating from the equilibrium position of one of the solvated solvent molecules.

The stiffness of the coil spring $k$ is determined from the condition of equality of two works $A_1 = A_2$, where $A_1$ is the work that is done by the moment of the forces arising from a deviation from the equilibrium position of one of solvent molecules solvated and oriented in the ion field, $A_2$ is the work that the returning force $F_2$ will perform, arising from spring deformation $\Delta l$. Deviation from the equilibrium position of one of the solvated molecules is schematically depicted in Fig. 1.3.

With a coordination number of 6 in a centrally symmetric field the cations are oriented and are within the first solvation radius of 6 solvent molecules. Solvation shell deformation leads to a deviation of one of the solvent molecules by an angle $\varphi$. In this case, the arising moment of forces will be $M = p \cdot (E \cdot \sin \varphi)$, where $p$ is the intrinsic dipole moment of the solvent molecule, $E$ is the value of the electric field strength formed by the central ion, within the first radius of solvation. The work committed by the moment of forces returning the solvent molecule to the equilibrium position $A_1 = M \cdot \varphi$. The work done by the returning force $F_2$ is $A_2 = F_2 \cdot d \cdot \varphi$, where $d$ is the length of the dipole formed by the solvent molecule. Moreover, $F_2 = k \cdot \Delta l = k \cdot (d \cdot \varphi)$.

The formula for calculating the stiffness value of a spiral spring in the SI units for the case when the solvent is water will be written as

$$k = \frac{pE}{d^2}\frac{\sin\varphi}{\varphi} \approx 6.33\cdot 10^{-5}\frac{\sin\varphi}{\varphi}, \text{ N/m.}$$

The solution of equation (1.6) gives an expression for the amplitude of the forced ion vibrations

$$b = \frac{F}{m\sqrt{\left(\omega_2^2 - \gamma^2\right)^2 + 4\lambda^2\gamma^2}}.$$

In the absence of friction of the solvated ion on the solvent molecule the amplitude of the forced oscillations of the ion will determine the expression

$$b_1 = \frac{F}{m\sqrt{\left(\omega_0^2 - \gamma^2\right)^2}},$$

Now one can determine the coefficient of decrease in the amplitude of ion vibrations due to friction:

$$k_f = (b_1/b)^{-1} = \left(1 + \frac{4\lambda^2\gamma^2}{\left(\omega_0^2 - \gamma^2\right)^2}\right)^{-1/2}.$$

The equation for the oscillations of the solvated ion $\Delta(t)$ with taking into account friction with the solvent molecules is written in the form

$$\Delta(t) = k_f\left(\frac{3A}{8gq_{ion}}(\sin 2\gamma t - 2\sin\gamma t) + C\right)^{=1/3},$$

where $g$ is the coefficient of proportionality in the Coulomb law in the SI system of physical quantities (see explanations for expression (1.5)). For the case when the solvent is water, friction on the solvent molecule of the solvated ion, the radius of the solvation shell of which is 0.2 μm, leads to a decrease in the amplitude of oscillations 1400 times, i.e. $k_f \approx 0.71 \cdot 10^{-3}$.

## 1.2. Experiment technique

The first chronologically conducted experiments were performed to study the action of an asymmetric electric field on the normal water LiCl solution (pH = 7). The following electrical field parameters were used. frequency 40 kHz, field strength in positive half-cycle 2 V/cm, asymmetry coefficient 0.75.

For simplicity, a periodic sinusoidal electric physical potential, for which the absolute values of the amplitudes of the half-periods are equal (Fig. 1.4), will be called symmetric. If the absolute values of the amplitudes of the half-periods differ they are asymmetric. The electric field between the potential and grounded electrodes (isolated from the solution) will be called symmetric and asymmetric, respectively.

Initially, the study of the processes accompanying the action of an external periodic electric field on salt solutions was carried out in homeotropic geometry: the direction of the electric field strength vector is perpendicular to the plane of the electrodes. The experimental setup includes a multi-section cell (Fig. 1.5). Overall dimensions of a cell: length: width: height 23:10:10 cm. Useful (internal) volume 735 ml. Cell consists of seven sections c1–c7 (Fig. 1.5, pos. 2; Fig. 1.6). Between sections there are potential grids of high transparency isolated from the solution (Fig. 1.5, item 4). As a material for manufacturing sections was selected vinyl plastic, which has a high electrical strength, inert in various environments (including aggressive ones) and having high technological properties. The material for side covers (Fig. 1.5, pos. 1) was 30 mm thick polymethylmethacrylate sheet. This cover thickness eliminates uneven section compression during assembly and thus possible leakage. The

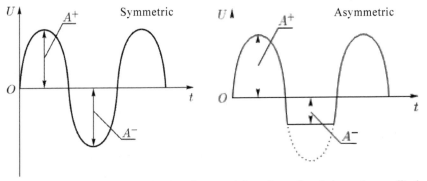

**Fig. 1.4.** The dependence of the electric potential on time: $A^+$ and $A^-$ are the amplitudes of the positive and negative half-periods, respectively.

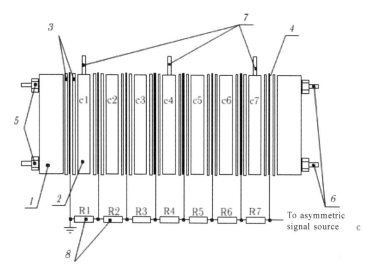

**Fig. 1.5.** The cell of 7 sections, No.1–No. 7.

joints of sections and potential nets were seales with rubber gaskets (Fig. 1.5, pos. 3), cut from sheet rubber 1 mm thick.

The cell is assembled using four steel studs (Fig. 1.5, pos. 6). Position 5 in Fig. 1.5 shows nuts of increased length for the possibility of a stronger tightening of the cell. In order to exclude the possibility electrical contact of metal studs with potential nets their main part located within the cell does not have grid threads or electrically insulation. In the extreme and central sections, pipes are made (Fig. 1.5, item 7) from chemical glass. They provide the supply and selection of the solution, as well as sampling for property analysis.

Pumping of solution is provided using a peristaltic pump NP – 70P–0.5.

Figure 1.6 presents the section – part of the cell. Position 1 is a technological hole for feeding and selecting the solution, immersion in a solution of analyzing probes, etc. Position 2 are holes for studs to assemble sections into a cell. The outer dimensions of the section: length 100 mm, width 100 mm, thickness 20 mm. Inner dimensions: 70 × 70 mm.

The experimental setup also includes an electrical part. Its functional purpose is the formation of an electrical signal with the necessary parameters and its supply to potential grids of the cells.

The main part of the electrical equipment of the installation is a device for generating an asymmetric high voltage signal. Such a

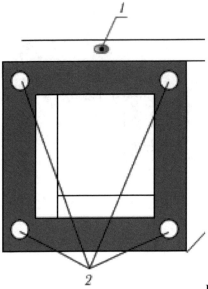

**Fig. 1.6.** Section.

device in mass execution could not be found, because of the special specificity of the required parameters. The problem was solved by the focused development of the required device.

The formation of various types of signals is possible using DAC and a computer, using its memory and service capabilities for preparing and storing the necessary signals in the digital form. The block diagram of the desired device is shown in Fig. 1.7.

The digital image stored in the memory of a personal computer (PC) can be converted into voltage using the analog output board (digital-to-analog converter, DAC). Then, the required amplitude of the signal is obtained using a voltage amplifier (VA).

To obtain an analog signal of the necessary form a device was assembled; the unified diagram of the device is shown in Fig. 1.8. The digital code of the next value of the generated signal which comes from the parallel port of the computer to the XP1 connector, is stored in the DD1 register and using the DAC DD2 and operating amplifier DA2 is converted to an analog signal. The R3C4 circuit suppresses high-frequency transients and smoothes the steps of the output voltage. Model voltage is applied to pin 15 of the microcircuit DD2 from the stabilizer, consisting of a zener diode VD1 and resistors R1 and R2. Trimmer R1 image value can be changed from 0 to $-9$ V, which regulates amplitude of the output signal. Chips D01 and DD2 supply voltage $+5$ V from the integral stabilizer DA1.

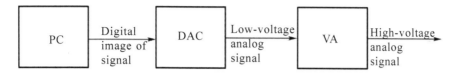

**Fig. 1.7.** Block diagram of the formation of a high voltage signal.

The maximum signal quantization frequency is mainly limited by the speed of the DAC and the parallel port of the computer. Used in the device DAC KR572PAA with the time of establishment in the output voltage of about 5 µs allows it to be brought up to 200 kHz. Thus, the upper limit of the frequency spectrum of the generated signals can be 50–100 kHz. Lower is not limited. If external process synchronization is required, the corresponding signal is supplied to terminal 12 (circuit PE) of XP1 connector. The computer can determine its logical level by analyzing the D5 bit of the code read from port 379H.

After receiving the required shape signal at the output of the device its amplitude increases to the required level. This problem is solved by using a voltage amplifier that provides increasing the amplitude of the input signal to 500 V.

Figure 1.9 shows a waveform of a signal with an amplitude of 150 V, frequency of 5 kHz and asymmetry coefficient 1 obtained using the above device: PC, DAC, voltage amplifier (VA). The oscillogram was obtained using the oscilloscope ASK-2031 and attached software. Amplitude sweep was 50 V/division (the standard divider was 1:10),

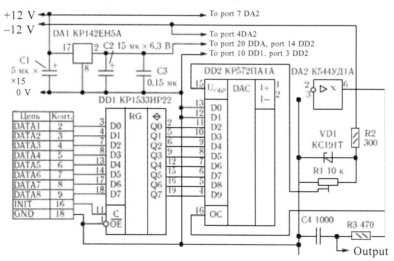

**Fig. 1.8.** Digital to analog conversion circuit.

time rotation 20 μs/div. The asymmetry coefficient implies a value equal to the ratio of the amplitudes of the half-periods of the signal: $k_{asym} = A^-/A^+$, where $A^+$, $A^-$ are the amplitudes of the positive and negative half-cycles, respectively. In this case, $k_{asym}$ can take a value from 0 (which indicates that the amplitude of the second half-cycle is zero) to 1 – the signal is symmetrical. Time of the rise of this signal is about 1 μs.

The distortion of the signal at the amplitude value is due to the features of the applied elements of the amplifier circuit. Measurements show that they will decrease with increasing signal frequency. The change in the asymmetry coefficient does not affect the value of the distortions. Figure 1.10 shows the waveform of an asymmetric signal ($k_{asym} = 0.5$); positive half-amplitude 200 V, frequency 50 kHz. Rise time in this case is 0.5 μs.

Figure 1.11 presents a waveform of a symmetric signal, whose amplitude is 500 V and frequency 50 kHz. With increasing voltage, the rise time remains the same (0.5 μs), but distortion is noticeably reduced.

The supply of an electrical signal to potential grids is carried out by the resistive divider R1–R7 (Fig. 1.5, pos 8).

To set the necessary signal parameters, the Aktakom oscilloscope ASK–2031 is used.. This appliance has enough a set of functions – frequency band 0–30 MHz sufficient for the experiments, 40 V maximum input, digital data storage, output of the data to print or to a PC via the RS-232 port.

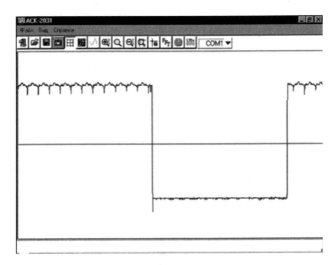

**Fig. 1.9.** Oscillogram of a symmetric signal: frequency 5 kHz, amplitude 150 V.

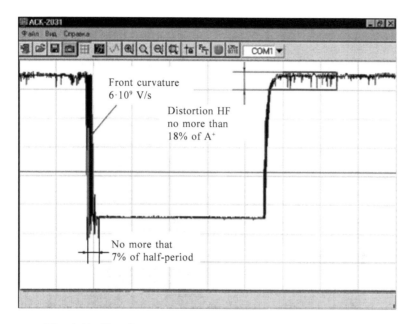

**Fig. 1.10**. Waveform: $A^+ = 200$ V, $k_{asim} = 0.5$, frequency 50 kHz.

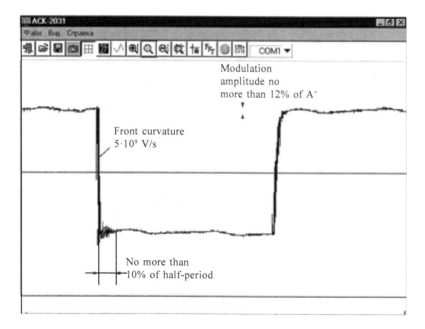

**Fig. 1.11**. Waveform: $A^+ = 500$ V, $k_{asim} = 1$, frequency 50 kHz.

The distillate was prepared using an electric aqua distiller DE-4. The specific electrical conductivity (SEC) of the water produced by this device is approximately 5 μS/cm (SEC of the tap water of the order of 400 μS/cm). Such measurements are carried out using the Anion 4120 conductivity meter. The range of the SEC values measured by this device is from 0.001 up to 100 mS/cm.

Weighing salts for the preparation of the studied solutions was carried out on an electronic scale Shimadsu BX320H. The resolution of the reading when weighing with a windshield is 1 mg.

In preparing the solutions, a magnetic stirrer PE-6110 with the possibility of heating was used.

The process of selective drift in the asymmetric electric field was studied using a solution of a mixture of salts of elements, similar in chemical properties. It is also advisable to use salts formed by one acid and obtained in identical technological processes. So, for example, at the initial stage, the drift parameters of the solvated $Mg^{2+}$ and $Ca^{2+}$ cations were studied in an aqueous solution of a mixture of salts of $MgCl_2$ and $CaCl_2$. The use of elements of complicated chemical properties undoubtedly complicate the set task, but at the same time, allows to exclude from consideration the influence of differences in chemical properties on the process under study. The main difficulty is in the analysis of selected samples for subject of the quantitative content of one or another cation. In studies of a solution of a mixture of salts of $MgCl_2$ and $CaCl_2$ the method of analysis of selected samples was selected x-ray fluorescence analysis. This method allows one to estimate the quantitative content of an element by the intensity of the lines of the element in the characteristic spectrum.

During the experiments, electric current, formed by the potential difference of a solution located in different sections of the cell, was measured. Current measurements were taken every 20 minutes by quickly immersing and extracting inert section from the volume of the electrodes. The measurements used a Protek 506 multimeter.

Figures 1.12 and 1.13 show the time dependence of the current, induced by the potential difference of the solution in different sections of the cell, which in turn is caused by the asymmetric electric field. No circulation took place (pump off).

After 6–7 hours of action of the asymmetric electric field, the induced current value stabilizes at a value of $(1.8 \pm 0.2)$ μA. When the generator is turned off, the current stops, but the potential of the grid between the 5th and 6th sections relative to the grounded

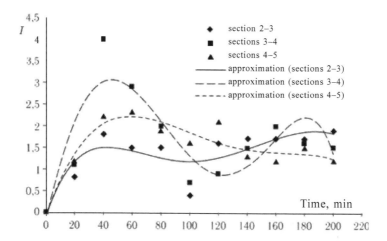

**Fig. 1.12.** Induced current induced by asymmetric electrical fields, µA ($E^+$ = 2 V/cm; $v$ = 40 kHz; asymmetry coefficient 0.75; approximation by 5th degree polynomial).

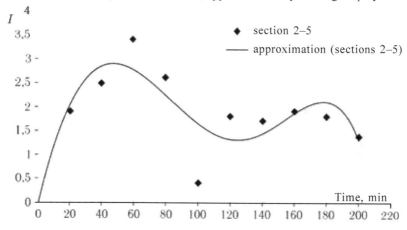

**Fig. 1.13.** Induced current induced by asymmetric electrical fields, µA ($E^+$ = 2 V/ cm; $v$ = 40 kHz; asymmetry coefficient 0.75; approximation by 5th degree polynomial simulation).

grid is (0.58 ± 0.02) V. After 18 hours, the potential disappears. In the moment of generator shutdown the sample from the 6th section stains phenolphthalein in raspberry red colour. After 18 hours, phenolphthalein does not stain – the solution becomes normal. With temperature decreasing by 10°C the accumulated excess concentration of the cationic aquacomplexes of the potential network is preserved within 24–26 hours. This is due to a decrease in the intensity of the chaotic thermal motion and, consequently, an increase in relaxation time.

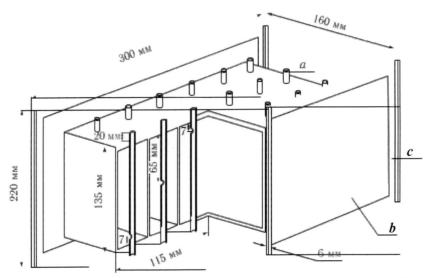

**Fig. 1.14.** Experimental setup: $a$ – nozzles for the selection of the solution; $b$ – a flat copper electrode; $c$ – electrode insulation (polymethyl methacrylate).

The experiments showed that under the influence of an asymmetric electric field, a drift of solvated cations occurs. When an asymmetric electric field is applied to the salt solution, selective drift of oppositely charged aquacomplexes: cationic and anionic, is induced In this case, there is observed separation of the drift directions: to the side of the ground electrode and to the side of the potential electrode. Thus, the electrophysical properties of the solution of electrodes differ: the normal solution of one of the electrode acquires alkaline properties from electrodes, and by the other one the acidic properties. The solution accumulates electrical energy.

A more complex experimental design was developed for the purpose of determining the spatial distribution of solvated cations with various inertial properties in the volume of a solution placed between two arrays by parallel flat electrodes. The experiments were carried out in the cell shown in Fig. 1.14. The experimental setup used (Fig. 1.14) consists of a set of identical sections (height 135 mm, length 150 mm, width 35 mm), placed between two potential electrodes.

Sections are separated by impermeable partitions. The electrodes representing a copper foil, are outside sections and are isolated by polymethyl methacrylate. Since one of section sizes is much smaller than the rest, in experiments the distribution of cations is analyzed only in the plane perpendicular to the electrodes (Fig. 1.14). Samples

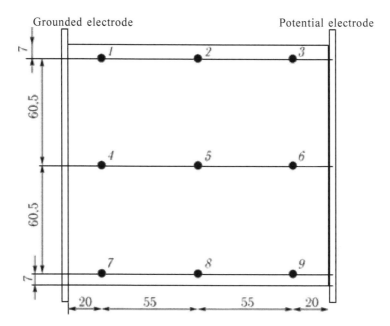

**Fig. 1.15.** Scheme of the experimental section.

from the volume of the experimental installations must be selected at nine points (see Fig. 1.15). To ensure that the disturbance introduced into the distribution of cations during selection samples is minimal, samples were taken from three sections operating under identical conditions: three from the top, three from the central parts and three from the bottom (see Fig. 1.15).

Sampling was carried out by gravity from rigidly fixed plastic capillaries (nozzles) placed in three different positions in each section. As a result, the parameters were determined of the distribution of metal cations in a plane perpendicular to the electrode plan for homeotropic geometry.

The experiments on the excitation of the phenomenon of selective drift of solvated ions under the action of an external 'asymmetric' electric field on the circulating salt solution were carried out on a set of devices and instruments. The block diagram of the complex is presented in Fig. 1.16.

The main source of voltage of variable frequency was a generator of sound signals GZ-109 (Fig. 1.16, item 1), the signal from which was received at the input of the RUB2250 preamplifier (2) with own power supply NES-100-12 (3). Then, the potential amplified by the transformer (4) through the forming device of an asymmetric

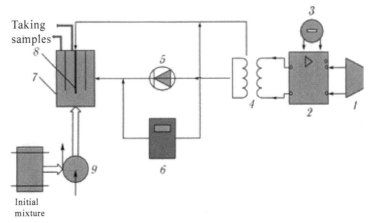

**Fig. 1.16.** The block diagram of the experimental complex.

waveform signal (5) was applied to the central electrode (7) of the experimental device. In this case, the second electrode (6) was grounded. During the experiments, the working solution from the tank (8) was supplied by the peristaltic pump (9) in the lower part of the experimental device.

The general view of the experimental device, which is a cylinder, inside of which the coaxially positioned main structural elements are shown in Fig. 1.17.

In the centre is a potential electrode (Fig. 1.17, item 1) from a steel bar, hermetically isolated from the working solution by a polyvinyl chloride tube. The grounded steel electrode of a cylindrical shape sealed against the working solution with a plastic wrap. The entire volume of working solution supplied in the lower part of the installation, is divided by a perforated polyethylene cylinder (5) into the inner and outer zones. This cylinder serves as a separator between the zones of enrichment and depletion of the solution with target ions.

Perforation of the cylinder (6) is made in the form of elongated vertical holes located azimuthally symmetrically along the cylinder forming line over its entire surface. The holes allow the ionic components of the working solution to move from one zone to another. The studied solution fed through the nipple (2) to the bottom of the unit the test moved upward in both zones with the same linear speed. The output of the solution at the top of the installation from each zone was carried out autonomously through nozzles (3, 4). These fittings were used to regulate the mass flow rate of the solution in the inner and outer zones. In this case, the speed of the

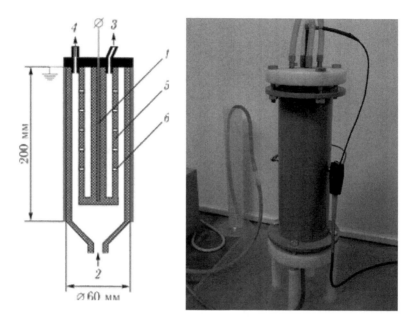

**Fig. 1.17.** Photo and schematic representation of a pilot plant in frontal section.

movement of the solution in the inner zone was 4–7% faster than the movement of the solution in the outer zone. In all experiments, the nature of the motion of the solutions corresponded to the laminar regime of fluid flow, and the linear velocity of the solution was not more than 4 cm/min.

Throughout the experiment, samples of solutions arising from both the internal and outer zones of the treated solution were taken. Analysis of samples for the content of magnesium and calcium ions was carried out at an atomic emission inductive coupled plasma spectrometer iCAP 6300 Duo.

The results of statistical processing of experimental data showed that the measurement error in all experiments did not exceed relative fractions of 0.014 of average values. It should be noted that using device 5 (Fig. 1.16) a positive wavelength of the initial sinusoidal signal formed and the volumes of the solutions of the outer and outer zones were exposed only to a positive potential sinusoidal field. This provided a jump-like movement of positively charged solvated cations to the grounded electrode with zero potential from the potential positively charged electrode (Fig. 1.17, item 1).

The main task was to detect a change in the concentration of magnesium and calcium cations in solutions of the internal and external zones. Such a fact would indicate the selective transfer of one type of cation from the outer zone to the inner one. In order to provide compelled transition of cations from the outer zone to the inner, the linear growth rate of the solution in it should be slightly higher. A similar mode of movement of the solution was established using clamps in each of the experiments. Since it was previously discovered that the effects of changes in the ion concentrations in the dilute solutions upon application asymmetric electric fields on them were observed in the range of frequencies of hundreds of hertz. Given this circumstance, in this work srudied were conducted in the frequency range from 10 to 1500 Hz.

## 1.3. Electroinduced drift of solvated calcium and magnesium ions

One of the steps to study the drift parameters of cationic aquacomplexes in an asymmetric electric field was experiments with solutions of a mixture of $CaCl_2$ and $MgCl_2$ salts

The sample composition was determined by X-ray fluorescent diffraction (XRFD) carried out in a setup in which the characteristic spectrum was excited by the bremsstrahlung of a chromium anode and with a planar semiconductor Si(Li) detector BDER KA 7K. In this case, the X-ray tube mode was used: 20 kV, 100 μA.

Using XRFD, the samples of the solutions were studied after experiments with parameters:

- concentration of $CaCl_2$, g/l　　　　　2.0,
- concentration of $MgCl_2$, g/l　　　　　2.0,
- frequency of the electric field, kHz　　5
- ambient envirenment temperature, ∘C　22
- pressure, mmHg　　　　　　　　　　247–252,
- amplitude of the first (positive)
half-cycle, V/cm　　　　　　　　　　5.7–14.5,
- electrical asymmetry coefficient signal 0; 0.5,
- exposure time to the electric field, h　　　6

Reading and processing of the obtained spectra were performed using SPECTR7 programs. An example of processing one of the spectra is shown in Fig. 1.18.

The intensity of the line (peak) of an element is directly proportional to its concentration. Seeing the possibility of changing the parameters of the XRFD equipment during measurementsm caused by many factors, the intensity of the calcium and magnesium

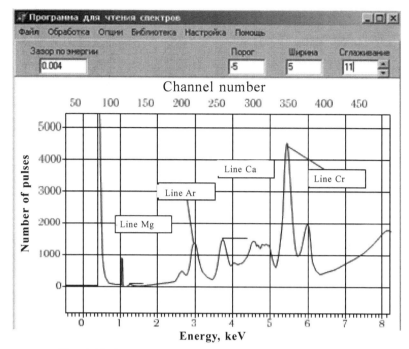

**Fig. 1.18.** Spectrum reading example using SPECTR7.

lines were normalized to the intensity pf argon lines. It is assumed that the argon content in air during the time of the measurements is constant (0.93%).

Thus, the concentration of magnesium ions in relative units is determined by the relation $c_{Mg} = n_{Mg}/n_{Ar}$, where $n_{Mg}$ is the intensity of the magnesium lines, $n_{Ar}$ is the intensity of the argon line.

Sampling was carried out after 6 hours of the effect of an asymmetric high-frequency field per aqueous solution of $MgCl_2$ and $CaCl_2$ salts. The experiments performed showed the following.

With an amplitude of the field strength in the first (positive) half-period of 5.7 V/cm and an asymmetry coefficient of 0 (the amplitude of the second half-period is zero) – experiment 1 – there is an oriented drift of solvated $Mg^{2+}$ and $Ca^{2+}$ cations from the potential to the grounded electrodes with predominant excitation of the drift of the cations $Mg^{2+}$: $C_{Mg}/C_{Ca}$ (centre) $\approx 1.232/1.056$. The results of processing the spectra are presented in Table 1.1. The range of the initial solution (Fig. 1.19) shows that the intensities of the lines of argon and calcium are approximately at the same level, the intensity of the magnesium line is ~1/6 of the intensity of the argon line.

**Table 1.1.** Relative concentrations of Mg and Ca ($A^+$ = 5.7 V / cm, $k_{asym}$ = 0)

| Position of sampling point | $C_{Mg} = c_{Mg}/c_{Mg}^{in}$ | $C_{Ca} = c_{Ca}/c_{Ca}^{in}$ |
|---|---|---|
| Section No. 1 | 0.172/0.168 | 1.047/0.966 |
| Section No. 4 | 0.207/0.168 | 1.020/0.966 |
| Section No. 7 | 0.139/0.168 | 1.794/0.966 |

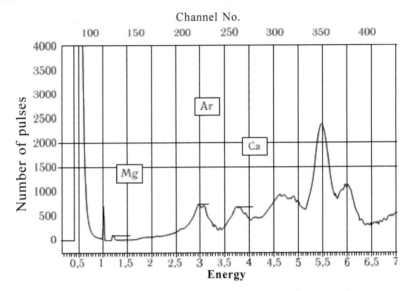

**Fig. 1.19.** Spectrum of the initial solution for the first experiment.

After 6 hours of field action, the spectrum analysis (Fig. 1.20) of the sample from section No. 1 shows an increase in the intensities of the magnesium and calcium lines relative to the argon line. The analysis of the spectrum from section No. 4 (Fig. 1.21) shows an increase in the intensity of the magnesium line and a decrease in the calcium line compared to the sample from section number 1. In the sample from section No. 7, the intensities of the lines of calcium and magnesium become less in comparison with the initial solution (Fig. 1.22).

With a voltage amplitude in the first half-period of 6.5 V/cm and an asymmetry coefficient of 0.5, an oriented drift of solvated $Mg^{2+}$ and $Ca^{2+}$ cations from the central sections of the separation cell toward the grounded and potential electrodes was observed. Figures 1.23–1.26 show the spectra of samples of the initial solution, as well as samples from sections with numbers 1, 4, 7 after the action of an electric field.

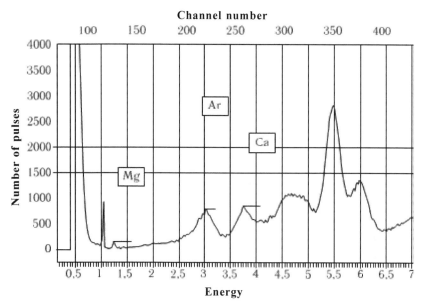

**Fig. 1.20.** The spectrum of the sample from section No. 1 after the action of the field ($A^+$ = 5.7 V / cm, $k_{asym}$ = 0).

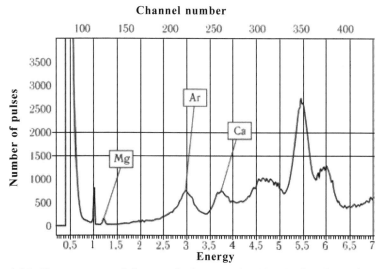

**Fig. 1.21.** The spectrum of the sample from section No. 4 after the action of the field ($A^+$ = 5.7 V/cm, $k_{asym}$ = 0).

All solutions were prepared from reagents of one batch, one at a time scheme. It is seen that the spectrum of the initial solution for the second experiment (Fig. 1.23) differs from the spectrum of the initial solution represented in Fig. 1.19. This suggests that there is

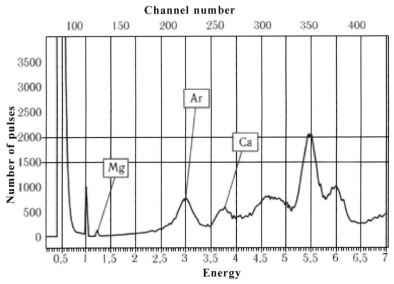

**Fig. 1.22.** The spectrum of the sample from section No. 7 after the action of the field ($A^+ = 5.7$ V/cm, $k_{asim} = 0$)

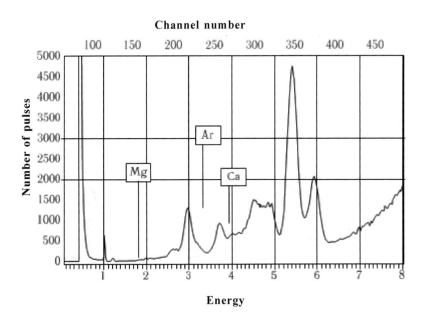

**Fig. 1.23.** The spectrum of the initial solution for the second experiment.

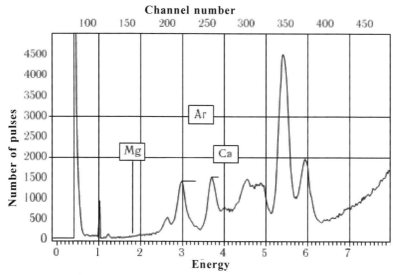

**Fig. 1.24.** The spectrum of the sample from section No. 1 after the action of the field ($A^+$ = 6.5 V/cm, $k_{asym}$ = 0.5).

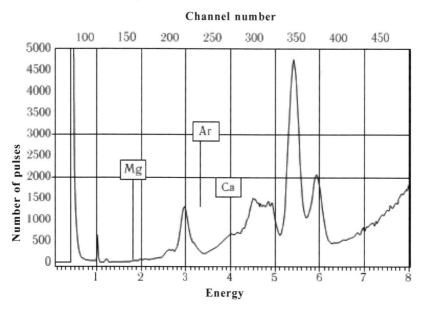

**Fig. 1.25.** The spectrum of the sample from section No. 4 after the action of the field ($A^+$ = 6.5 V/cm, $k_{asym}$ = 0.5).

a change in parameters of the XRFD installations. It is seen that the intensity of the calcium line in the spectrum of the initial solution is lower than the intensity of the argon line.

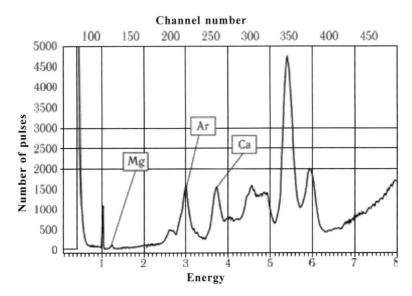

**Fig. 1.26.** The spectrum of the sample from section No. 7 after the action of the field (A + = 6.5 V / cm, $k_{asym}$ = 0.5).

After the action of the field, the intensity of the calcium line in the spectrum of the sample from section No. 1 (Fig. 1.24) exceeds the intensity of the argon line, the intensity of the magnesium line also increases. Spectrum analysis samples from section No. 4 (Fig. 1.25) shows a decrease in the content of magnesium and the inability to determine the amount of calcium using this installation (there is no peak of calcium).

The spectrum of the sample from section No. 7 (Fig. 1.26) is characterized by a smaller (relative to the initial solution) increase in the magnesium content and a slightly larger increase in calcium. Experimental data indicate the predominant excitation drift of $Ca^{2+}$ cations. The grounded electrode has a $C_{Ca}/C_{Mg}$ concentration ratio ≈ 1.516/1.213, at the potential electrode ≈1.539 /1131 (Table 1.2).

The spectrum of the sample solution for the third experiment shows pronounced lines of calcium and magnesium; a less pronounced argon line (Fig. 1.27) with its constant amplitude during the X-ray diffraction (Figs. 1.27–1.30).

The increase in the amplitude of the tension to 14.5 V/cm at the same asymmetry coefficient leads to the establishment of the drift of aquacomplexes from the grounded to the potential electrode (Figs. 1.27–1.30).

**Table 1.2**. Relative concentrations of Mg and Ca ($A^+$ = 14.5 V / cm, $k_{asym}$ = 0.5).

| Position of sampling point | $C_{Mg} = c_{Mg}/c_{Mg}^{in}$ | $C_{Ca} = c_{Ca}/c_{Ca}^{in}$ |
|---|---|---|
| Section No. 1 | 0.299/0.533 | 1.48/0.966 |
| Section No. 4 | 0.58/0.533 | 3.86/3.29 |
| Section No. 7 | 0.72/0.533 | 4.53/3.29 |

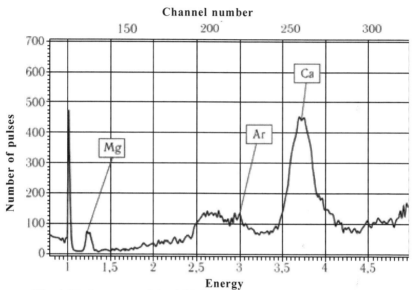

**Fig. 1.27**. Spectrum of the initial solution for the third experiment.

The increase of the amplitude of the strength to 14.5 V/cmat the same asymmetry coefficient leads to the establishment of the drift of the aquacomplexes from the grounded to potential electrode (Fig. 1.27–1.30) with predominant excitation of the drift of Ca$^{2+}$ cations. At the grounded electrode, the Ca and Mg concentrations decrease, and they increase in the central cell (Table 1.3), forming the ratio $C_{Ca}/C_{Mg} \approx 1.173 / 1.088$, and the potential electrode: $C_{Ca}/C_{Mg} \approx 1.377/1.351$. Analysis of the spectrum of the sample from section No. 1 (Fig. 1.28) shows a significant (approximately two-fold) decrease of the magnesium content and even more significant reduction of the calcium content. According to the spectrophot of the robe from section No. 4 (Fig. 1.29) it can be seen that the intensity of the magnesium line has increased and the magnesium content now exceeds the magnesium content in the initial solution. The same is observed for calcium, but is more pronounced.

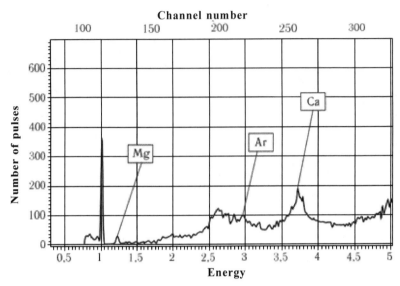

**Fig. 1.28.** The spectrum of the sample from section No. 1 after the action of the field ($A^+$ = 14.5 V/cm, $k_{asym}$ = 0.5).

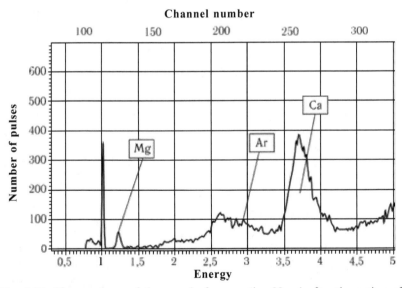

**Fig. 1.29.** The spectrum of the sample from section No. 4 after the action of the field ($A^+$ = 14.5 V/cm, $k_{asym}$ = 0.5).

Figure 1.30 presents a range of samples from section No. 7. Intensity of the strength of the calcium line exceeds the corresponding value for the initial solution by 1.377 times. For magnesium, this value is 1.351.

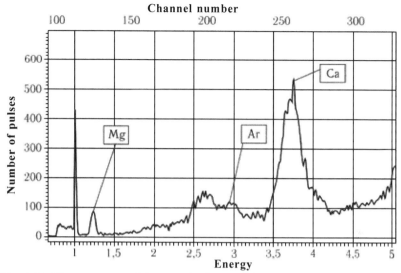

**Fig. 1.30.** The spectrum of the sample from section No. 7 after the action of the field ($A^+$ = 14.5 V/cm, $k_{asym}$ = 0.5).

**Table 1.3.** Relative concentrations of Mg and Ca ($A^+$ = 14.5 V/cm, $k_{asym}$ = 0.5).

| Position of sampling point | $C_{Mg} = c_{Mg}/c_{Mg}^{in}$ | $C_{Ca} = c_{Ca}/c_{Ca}^{in}$ |
|---|---|---|
| Section No. 1 | 0.299/0.533 | 1.48/3.29 |
| Section No. 4 | 0.58/0.533 | 3.86/3.29 |
| Section No. 7 | 0.72/0.533 | 4.53/3.29 |

Thus, we can formulate an unambiguous conclusion that in an aqueous solution of a mixture of salts of $CaCl_2$ and $MgCl_2$ takes place the effect of oriented, selective drift of cationic aquacomplexes upon exposure to an asymmetric electric field with a frequency of 5 kHz.

## 1.4. Electroinduced drift of solvated cerium and lead cations

To conduct experiments to study the process of selective drift, we used a solution of a mixture of $Ce(NO_3)_3$ and $Pb(NO_3)_2$ salts with concentrations of metal salts of 0.1 g/l.

Spectrophotometric analysis was chosen as a method for quantifying the changes in solution concentration. This method allows to evaluate the content of an element in a solution by the intensity of the absorption lines in the characteristic spectrum.

The analysis of the samples was carried out on a spectrophotometer Evolution UV600 with a spectral range in the ultraviolet region. This spectrophotometer is designed to measure spectral transmittances of liquid and solid substances in the spectral region from 190 to 1100 nm, as well as for measuring the diffuse or specular reflection coefficients of flat objects.

The error in setting the wavelength was not more than 0.03 nm; reproduction error not more than ±0.1 nm; photometric range: optical density (A) was 0.3–4 A; research speeds 0.05; 0.1; 0.2; 1; 2; 5 and 10 nm/s.

Reading and processing of the obtained spectra was performed using VISIONpro programs. An example of processing one of the electronic spectra is shown in Fig. 1.31.

The curves for solutions with a known concentration were constructed using a calibration graph, based on which the concentration of cerium cations was determined. The type of graphs taken for calibration is shown in Fig. 1.32.

Using a UV spectrometer, samples of the solutions were investigated. after experimenting with the parameters:

• Ce concentration, g/l                                             0.1,

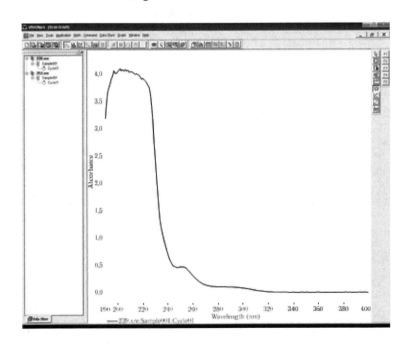

**Fig. 1.31.** Example of recording the electronic absorption spectrum of $Ce(NO_3)_3$ using the VISIONpro program

- concentration of Pb, g/l     0.1,
- frequency of the electric field, Hz     100 and 200,
- ambient temperature of medium, °C     22,
- pressure, mmHg     747–752,
- the amplitude of the first (positive) half-period, V/cm     78.9–98.7,
- asymmetry coefficient of electrical signal     0.2; 0.5,
- exposure time of electrical field     2 h 15 min; 4 h 30 min; 8 h; 20 h

The experimental setup includes a three-section cell (Fig. 1.33). The overall dimensions of a cell (length: width: height) 102: 10: 10 cm. Useful (internal) volume 320 ml. The cell consists of three sections c1–c3 (Fig. 1.33, item 2).

Sampling was carried out after 2 h 15 min; 4 h 30 min; 8 h; 20 hours of action of an asymmetric high-frequency field on an aqueous solution Ce $(NO_3)_3$ and $Pb(NO_3)_2$.

In numerical experiments, it was shown that in the case of effects on the salt solution of an electric field with equal amplitudesof the tension in the positive and negative half periods('symmetric field') selective drift of solvated ions does not get excited. Only in the

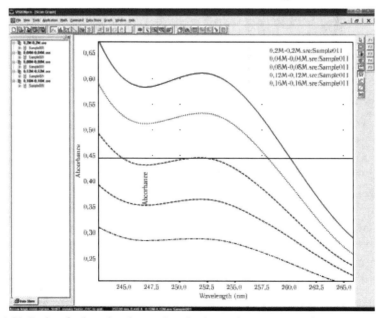

**Fig. 1.32.** Peaks of solvated cerium cations in electronic spectra at various concentrations of Ce $(NO_3)_3$ in solutions.

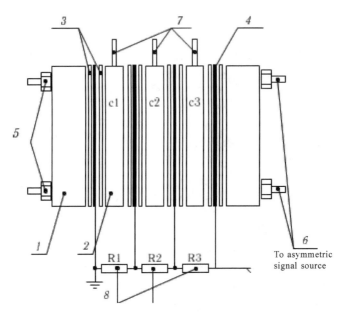

**Fig. 1.33.** Experimental cell: 3 sections No. 1–No. 3.

presence of asymmetry of the electrical signal the effect of selective drift should be expected.

When a 'symmetric' field was applied to the solution, that is, with an asymmetry coefficient of the electric signal equal to unity, the concentration changes in the sections did not exceed 0.4% (Fig. 1.34). These minor changes can be explained by the measurement error.

When exposed to a solution of the field with a frequency of 200 Hz and an asymmetry coefficient of 0.5, with an amplitude of the positive half-cycle of 600 V there is a tendency to increase the concentration of solvated cerium cations in the first and third sections when the solution is depleted in the central (second) section. Separation coefficient (during selection) in the first section) is 1.015. When the field is applied for 4 hours 30 min the trend is unstable.

With a decrease in frequency to 100 Hz, with the remaining field parameters unchanged, a tendency to increase the concentration of the grounded electrode becomes more distinct. In the first section there is observed a noticeable increase in the concentration of cerium cations, as well as a less significant growth is noticeable in the second section, when the solution is depleted of solvated cations in the third section.

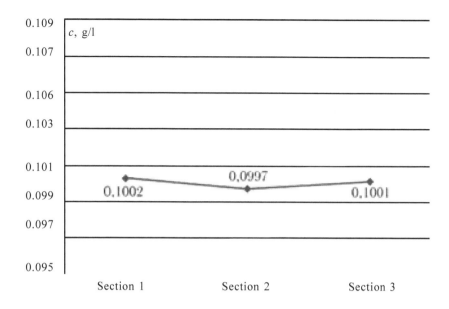

**Fig. 1.34.** The concentration of cerium cations $c$ (g/l) in the experimental cell after exposure to a field of 100 Hz, 450/450 V for 4 hours 30 minutes.

In this case, the duration of the field exposure of 4 h 30 min is close to the optimal start, since the separation coefficient (during in the first section) at this moment is 1.054 (Fig. 1.35), and at increasing the duration of the field to 8 hours separation coefficient reduced to 1.026. Reducing the exposure time of the field to 2 hours 15 minutes also reduces the separation coefficient to 1.0166.

## 1.5. Induced redistribution of solvated cations of cerium and nickel in water of their chloride solution

The experimental setup used is presented on Figs. 1.14 and 1.15. The parameters of the experiments:
- concentration of $CeCl_3$, g/l        1,
- concentration of $NiCl_2$, g/l        5,
- frequency of the electric field, Hz    100 and 200,
- ambient temperature of medium,°C    22,
- the amplitude of the field strength in solid half-cycle, V/cm        56,
- the ratio of the amplitude of the negative half-period to positive period amplitude    750/150; 600/300,
- time of action of electric field, h        0.5; 1; 2.

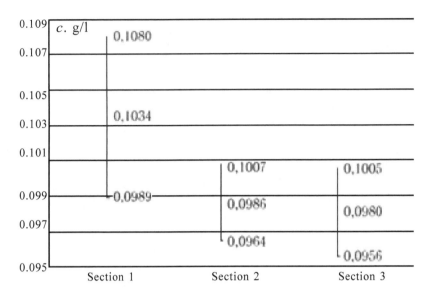

**Fig. 1.35.** The concentration of cerium cations $c$ (g /l) in the experimental cell after exposure to a field of 100 Hz, 600/300 V for 4 hours 30 minutes.

The choice of elements and concentration values are due to the features of quantitative analysis on spectrophotometric equipment.

Duration of exposure to an electric field with a frequency of 100 Hz in the experiments was 30 min, 1 and 2 hours. Each experiment was performed three times.

Reading and processing of the obtained spectra was performed using the VISIONpro program. From the curves for solutions with a known concentration, a calibration graph was constructed based on which the concentrations of metal cations were determined.

### 1.5.1. The results of the experiments (without circulation of the solution)

Summary charts of experimental results after an hour-long exposure to an electric field with a frequency of 100 Hz and a strength of 750/150 V per solution of cerium and nickel chlorides are presented in Figs. 1.36 and 1.37. Concentration values are presented as a percentage of the value of the concentration of the stock solution.

An analysis of the results shows that the drift of cations of cerium and nickel have a similar character and are directed mainly down to the bottom of the section, drift intensity in the direction there are fewer electrodes. In general, there is a decrease in concentrations of cations (both cerium and nickel) in the central part of the section and their increase near electrodes, more significant potential

While maintaining the parameters of the electric field and reducing the exposure time up to thirty minutes there is a decrease in the concentration of both cerium and nickel in the central part of the section and, accordingly, an increase in concentration near the electrodes, more significant up potential. A distinct, as in the previous experiment, increase in the concentration of solvated cations in the bottom of the section is not observed. The reproducibility of he cation concentration distribution parameters with decreasing duration of the field exposure falls.

With an increase in exposure time to two hours, accumulation of the solvated cations in the upper part of the potential electrode section and in the bottom part of the ground electrode takes place The drift is mainly directed towards the bottom of the section. Distributions of the solvated cations under a two-hour exposure (as well as during a half-hour) are reproduced worse – stability of the process of separation of solvated cations is low.

Based on the results of the experiments, the points of optimal selection, dump and supply of the solution for the organization of the process of separation of cerium and nickel cations were determined.

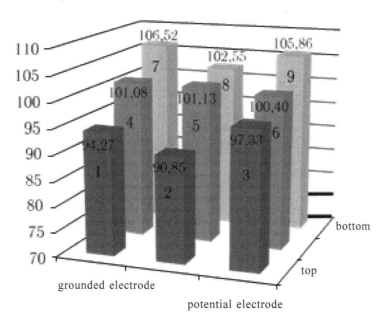

**Fig. 1.36.** The distribution of nickel cations (as a percentage of the initial value of 100%).

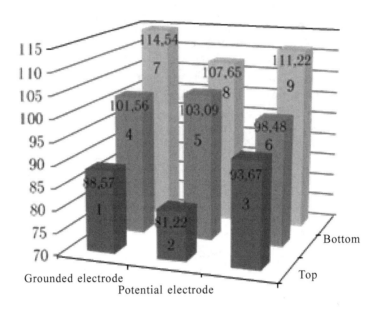

**Fig. 1.37.** Distribution of cerium cations (as a percentage of the initial value of 100%) electrode. The error in determining the concentration of metal cations does not exceed 9%.

The concentrations of the components in the selection are equal: $Ce^{3+}$–$C_i^i$; $Ni^{2+}$– $C_j^i$, and in the dump – $C_i^"$ and $C_j^"$, respectively. In this case, the selection is enriched in $Ce^{3+}$ cations and depleted in $Ni^{2+}$ cations, but the dumping is vice versa: it is depleted in $Ce^{3+}$ cations and enriched in $Ni^{2+}$ cations.

The maximum separation coefficient can be obtained if the dumping is represented by point 9 (see Figs. 1.14, 1.15 – the bottom of the section at the potential electrode), behind the dumping – point 2 (central upper part of the section). Thus, after an hourly exposure of the field of 750/150 V with a frequency of 100 Hz to the aqueous solution of the mixture of the salts of $CeCl_3$ and $NiCl_2$ the concentration values in the absence of solution circulation through the section are: $C_i' = 1.1454$; $C_j' = 1.1065$; $C_i" = 0.8122$; $.C_j" = 0.9085$. In this case, relative concentrations of cations or cerium and yttrium are: $R_{ij} = 1.00000$ (initial solution); $R_{ij}' = C_i'/C_j' = 1.0755$ (selection) and $R_{ij}" = C_i"/C_j = 0.8940$ (dumping).

The separation coefficients without circulation of the solution are:
- in the selection $\alpha_{ij} = R_{ij}'/R_{ij} = 1.0755$;
- in the dumping $\beta_{ij} = R_{ij}/R_{ij}" = 1.1186$;
- full $q_{ij} = \alpha_{ij}-\beta_{ij} = 1.2027$.

## 1.5.2. The results of the experiments (with circulation of the solution)

Based on data from experiments with a motionless solution, a series of experiments was conducted with circulation of the solution through one section of the experimental setup. Saline pumping was provided using a peristaltic pump, the solution was supplied to the centre of the section (point 5 in Fig. 1.15), selection was carried out from points 1, 3, 7, 9 in Fig. 1.15. Experiments were carried out with an asymmetric electric field voltage of 750/150 V, frequency 100 Hz.

The results of the experiments (Fig. 1.38) are in good agreement with the results obtained in experiments with a stationary solution. An increase in the concentration of the solvated cations (ascerium and nickel) is observed in the bottom of the section in the region of the grounded electrode (point 7 in Fig. 1.15). A decrease in the concentration of cations occurs on the potential electrode in the upper parts of the section (point 3 in Fig. 1.15). Thus, when circulatingthe solution at a speed of 7 l/h the main direction of drift of the solvated cations (both cerium and nickel). Relative change in the concentration of solvated $Ce^{3+}$ cations in the selection significantly exceeds the change in the concentration of $Ni^{2+}$ cations.

With a decrease in the circulation rate of the solution to 1.7 l / h, the total drift direction of cationic aquacomplexes remained, but its intensity decreased (Fig. 1.39).

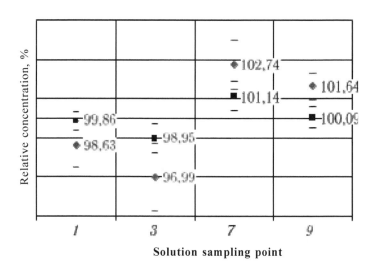

**Fig. 1.38.** Distribution of cerium and nickel cations at a circulation rate of the solution ≈ 7 liters/hour.

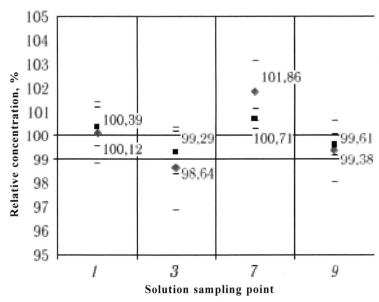

**Fig. 1.39.** Distribution of cerium and nickel cations at a solution circulation rate of ≈ 1.7 liters/hour.

With an increase in the speed of pumping the solution to 14 liters per hour the overall distribution pattern of the solvated cations has changed somewhat. The maximum concentration value is observed at sampling point 7 in Fig. 1.15, and the minimum value has shifted from point 3 to point 9 in Fig. 1.15. At a given pumping rate, the variation in the concentration values in the experiments increased.

From the values obtained in the experiments, the values of the separation coefficients were calculated. For 'selection' point 7 in Fig. 1.15, in which the maximum difference in the concentrations of cations of cerium and nickel was observed was taken. The values of the separation coefficients during the selection of the 'dumping' solution at various points are given in Table 1.4.

The values given in Table 1.4 values suggest that stable separation of $Ce^{3+}$ and $Ni^{2+}$ cations can be achieved at a circulation rate of 7

**Table 1.4.** Values of separation coefficients of solvated cations $Ce^{3}$ and $Ni^{2+}$under the effect of asymmetric field (100 Hz, 750/150 V)

| Position of sampling point | Dumping at point 1 | Dumping at point 3 | Dumping at point 9 |
|---|---|---|---|
| 14 | 1.021±0.036 | 1.019±0.051 | 1.034±0.051 |
| 7 | 1.029±0.019 | 1.0236±0.025 | 1.000±0.017 |
| 1.7 | 1.014±0.021 | 1.018±0.034 | 1.014±0.019 |

±

l/h, providing selection from point 7 (the bottom of the section of the ground electrode), and the dump from points 1 or 3 in Fig. 1.15 The maximum value of the separation coefficient is ensured when selecting the waste solution at point 3.

The maximum separation coefficient of the solvated cerium and nickel cations in the case of an aqueous solution of their chlorides was achieved in experiments under the action of an electric field of voltage 750/150 V with a frequency of 100 Hz per aqueous solution for an hour.

## 1.6. Separation of solvated calcium and magnesium cations by the action of an external periodic electric field and a moving solution

Figure 1.40 shows the nature of the change in the concentration of calcium ions in the internal (curve 1) and external (curve 2) zones in the frequency range from 20 to 180 Hz [8]. The experiments were performed on the setup shown in Fig. 1.17. In the following sections, it will be shown that the initiation of the phenomenon of electroinduced drift should be expected at frequencies of tens of Hz, when a self-consistent field is formed in the solution, and the size of the solvated ion (cluster) formed by the ion and solvent molecules forming the solvate shell is inversely proportional to the square root of of the salt concentration in the solvent. Frequency values, in turn, are inversely proportional to the value of the moment of inertia of the cluster.]

As can be seen, in the range of 80–120 Hz there is a noticeable increase in the content of calcium ions in the inner zone of the working solution with a distinct maximum. It is important to note the corresponding decrease in their concentration in the outer zone, that is, it has a redistribution of the concentration of one of the ionic components between the interstices of the working solution takes place, which essentially means the process of their separation. It is likely that in this frequency range there is a deformation of the solvate shells of calcium ions, and possibly even static destruction of the outer layers of solvation, where the binding energy of the solvate water molecules with the central ion is minimal in comparison with the water molecules that make up the primary solvation shells. This circumstance creates the conditions for the directed transfer of desolvated ions in a potential field, which, in accordance with the terminology adopted earlier, determines the effect of their selective

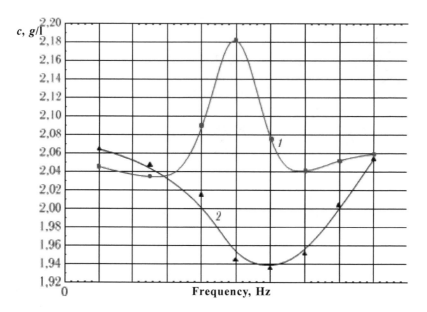

**Fig. 1.40.** The influence of the frequency on the content of calcium ions in the separation zones at a voltage of 285 V (field strength in the cell 86.4 V/cm) and the initial concentration of 2.065 g/l. The circles indicate the concentration in the inner zone, the triangles indicate the concentration in the outer zone.

electro-induced drift. Since every electrolyte solution has the property of electroneutrality, then any change in the local volume of its electric charge – positive or negative – should lead to instantaneous compensation of this charge by moving ions of the corresponding type into this volume. So, if in our case there is a transfer of calcium ions from the outer separation zone to the inner one, creating in the latter an excess content of ions of a positive charge, then an equivalent amount of magnesium ions should be transferred from the inner zone to the outer one. This phenomenon can be interpreted both as a process of electromigration migration and as an exchange process between different classes of ions. Thus, in the inner zone there should be a decrease in the content of magnesium ions, and in the outer zone, on the contrary, an increase in the concentration of these ions. Indeed, if we turn to Fig. 1.41 then it becomes obvious that the change in the concentration of magnesium ions with respect to the content of calcium ions in the inner and outer separation zones in the indicated frequency range is clearly expressed inverse.

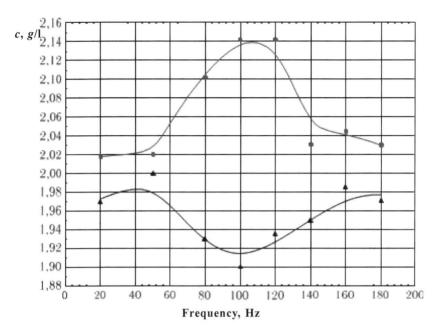

**Fig. 1.41.** The change in the concentration of magnesium ions in the separation zones at a voltage of 285 V (field strength in the cell 86.4 V/cm) and the initial content of 1.986 *g/l*. The circles indicate the concentration in the inner zone, triangles – concentration in the outer zone.

In its own way, the frequency dependence of the content of calcium ions is resonant in nature, that is, the concentration extremes for these ions are determined in a rather narrow frequency range.

The same can be said about the frequency dependence of the content of magnesium ions (Fig. 1.42). However, extreme concentrations of magnesium ions in the separation zones are observed in the region of more high frequencies, that is, when the energy of the external field exceeds the threshold values of the resonance phenomena for the same calcium ions. This is apparently due to the difference in the bond length of the solvation shell with the central ion. The radius of the outer electron shell of a magnesium ion is smaller than that of calcium ion; therefore, the coulomb interaction of the magnesium ion with water molecules that make up the solvate shells forms a shorter bond, which corresponds to a higher natural frequency of the bond. Cations have only the coordinating effect on the solvent molecules in the first and second solvate spheres, and the number of solvent molecules in the solvate shell is determined by the screening radius of the cation charge by the total charge of polarized solvent molecules.

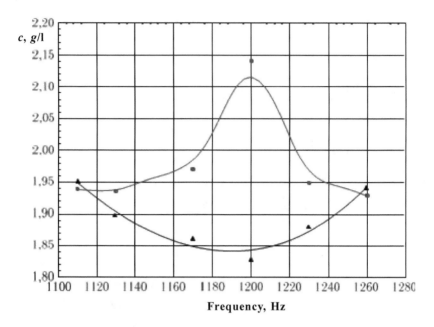

**Fig. 1.42.** The frequency dependence of the content of magnesium ions in the separation zones at a voltage of 285 V (field strength in the cell 86.4 V/cm) and the initial concentration of 1.931 $g/l$. The circles indicate the concentration in the inner zone, the triangles indicate the concentration in the outer zone.

The following sections will show that the directional drift of the solvated ions is also excited at frequencies corresponding to various components of the rotational–translational motion of the ion-solvation shell system, and at frequencies corresponding to transitions of oscillatory movements into rotational ones wherein the frequency values are units of kHz. The resonance can be expected:

• at the frequency of cooperative rotational motion of $H_2O$ molecules combined into a solvation shell relative to an axis passing at a distance equal to the outer radius of the solvation shell from the energy centre;

• at the frequency of cooperative rotational motion of $H_2O$ molecules combined into a solvate shell relative to the centre of inertia of the solvated ion (cluster);

• at the frequency of rotational motion of the cluster as a whole;

• at the frequency of transition of the oscillatory motion to the rotational one.

Consequently, for the resonant deformation of the solvate shells of magnesium ions, a higher value of the external electric field energy will be required, which will increase with increasing frequency fluctuations. Theoretical ideas about the electroinduced drift of solvated ions indicate that one of the factors causing this phenomenon is the amplitude of the electric potential, which determines the electric fields. The experiments carried out at voltage values of 40 V (field strength in the cell 12.1 V/cm) showed that, within the limits of the measurement error, the concentrations of magnesium and calcium ions did not change and remained at the level of their initial values. Since potential is the main energy characteristic of an electric field, then, apparently, there is a certain threshold value of the electric field strength, which determines the energy minimum, which in a certain frequency range causes the deformation of the solvation shells of ions and, accordingly, their selective electro-induced drift. In the case under consideration the threshold value of the electric field lies in the range from 12 to 100 V/cm. In relation to the studied special character of separation, the total coefficients of the separation of cations of calcium and magnesium are determined by the measurement results. To talk about a single separation coefficient, complete information is needed on all possible mechanisms of selective mass transfer. The process of electroinduced drift at the interface between the outer and inner zones, which leads to the exchange of different-grade ions between these zones, is presumably prevailing. An additional driving force that provides, along with electromigration transfer, the movement of ions from the outer zone to the inner, can be represented by an excess hydrodynamic pressure, which occurs due to the difference in the speeds of the solutions in the inner and outer zones. In this regard, the question of how the velocity ratio affects the movement of solutions in zones on the effect of separation requires a separate consideration. The simultaneous mass transfer of a certain fraction of competing ions over a certain period of time across the border of the outer and inner zones over the entire area of their contact can, apparently, be interpreted as a single separation coefficient.

## 1.7. Magnetically induced mass transfer in salt solutions

The scheme for exciting intense mass transfer in a salt solution placed in an alternating magnetic field is quite simple and similar to a transformer circuit, the secondary winding of which is formed by a 'coil' of the solution (see Fig. 1.43). This 'coil' is located in a dielectric vessel (polymethyl methacrylate), placed in a variable magnetic field.

The field in the magnetic core of the transformer is excited by alternating current in a conventional primary winding. A feature of the circuit is that the 'turn' of the solution is 'open' by the diode or (which is the same) by the load of the 'turn' is a semiconductor junction. Only in this case, intense mass transfer is excited in the solution. The intensity of mass transfer suggests that the primary factor is the excitation of the motion of particles having a significant 'hydrodynamic radius'. Only in this case, their motion due to friction can cause the movement of surrounding particles and the entire solution as a whole. These particles are aquacomplexes – clusters having submicron sizes. Clusters are formed by both cations and anions and are charge neutral at time intervals exceeding the time between Brownian collisions of solvent molecules.

The experiments were carried out on the basis of a magnetic circuit (transformer core) assembled from sheet electrical steel. Its dimensions were $400 \times 200 \times 50$ mm. The primary winding contained 200 turns of copper wire with a diameter of 3 mm, the voltage on the primary winding was 150 V (50 Hz). The secondary winding was a polyethylene hose with a diameter of 20 mm, filled with a solution. In the case of an aqueous solution of $CuSO_4 + H_2SO_4$ electrolyte, the voltage at the ends of the 'open' secondary winding changed from 1 to 3.2 V with an increase in the number of coils of the solution from 1 to 6. The current strength in the secondary circuit (in solution), 'loaded' on the diode, reached 2.8 A with the

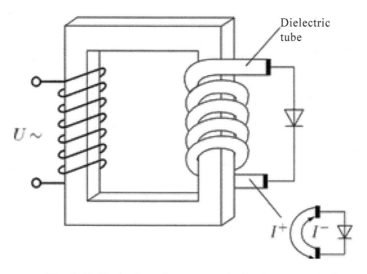

**Fig. 1.43.** Excitation of mass transfer by the magnetic field.

number of turns equal to 6. Figure 1.44 shows the current–voltage characteristic of the 'secondary' winding 'loaded' on the diode. A very important (experimentally established) fact is that the current strength clearly depends on the diameter of the hose forming the secondary winding. With large hose diameters (with other conditions being identical), the current strength is greater. At small diameters, the current could not be excited at all. This clearly indicates that the current is formed by the directional movement of solvated ions having orders of magnitude greater mass as compared to the mass of a 'bare' ion. The Larmor radius of such solvated ions – clusters is large and comparable to the diameter of the hose inside which the mass transfer is excited. For small hose diameters, the induced motion of the solvated ions (clusters) is hindered by the appearance of collisions with the walls of the hose.

Excitation of a salt solution in a dielectric with an alternating magnetic field of oppositely directed currents of anionic and cationic clusters makes it possible to organize two technological processes.

The first is the enrichment (purification) of the solution by anions or cations, that is, the separation of chemical elements. The second is electrolysis. In this case, at the ends of the 'open loop of the solution' it is necessary to place the electrodes, which should still be 'loaded' on the diode.

Electrolysis will occur at the interface between the electrodes and the solution. For example, in the case of an aqueous solution of $CuSO_4 + H_2SO_4$ electrolyte, intense copper evolution occurs on one of the graphite electrodes.

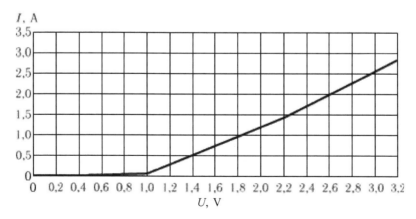

**Fig. 1.44.** Current–voltage characteristic of the 'secondary winding' from a solution of $CuSO_4$.

## Conclusions

The results of experiments on the effect of an asymmetric electric field on aqueous salt solutions show that when an asymmetric electric field is applied to a salt solution, selective drift of oppositely charged aquacomplexes is induced: cationic and anionic. In this case, a separation of the drift directions is observed: to the side of the ground and to the side of the potential electrodes. The solution accumulates electrical energy. When drifting in one direction for cationic (anionic) aquacomplexes, selection is observed due to the difference between the normal (with respect to the plane of the electrodes) components of the velocity vector of the centres of inertia. In an aqueous solution of a mixture of salts of $CaCl_2$ and $MgCl_2$ the effect of oriented, selective drift of cationic aquacomplexes occurs under the influence of an asymmetric electric field, the frequency of which does not exceed 5 kHz.

The ratio of the displacements due to the field-induced rotational-translational motion and the displacements caused by the chaotic motion of the solvated cations is determined by many parameters. Of these, field tension $E$ and solution temperature $T$ can be distinguished. The trajectory of the aquacomplex is determined by a set of at least 7 parameters: field strength; frequency asymmetry coefficient; bond lengths in a dipole (polarized aquacomplex); mass ratios of positive and negative parts of the dipole; masses of the aquacomplex as a whole; the polarization coefficient of the aquacomplex.

The performed experiments prove the possibility of using the discovered phenomenon of electroinduced selective drift of solvated ions in salt solutions under the action of the asymmetric electric field for organizing the technological process of enrichment of solutions by cations of the target metal.

It was found that the excitation of the effect of selective electroinduced drift of solvated ions can be observed in at least two frequency ranges. Within each of the intervals there are frequencies corresponding to eigenfrequencies of oscillations of the 'ion–solvate shell' system as a spherical rotator (rotation of the solvate shell relative to the central ion) or as a system of rigidly connected central ion and hollow shell.

The frequency dependence of the cation content in the solution of a mixture of their salts indicates that for each of the competing cations there is a certain frequency range in which the rate of directional drift of one of them increases significantly.

# 2

# Physics of the process of electro- and magnetoinduced mass transfer in salt solutions

An unexcited aquacomplex whose solvation shell is not deformed, is neutral. The shell performs the function of a screen which prevents the action of the constant component of the high-frequency electric field. There is a component entering in the canonical form of expansion in a Fourier series of a periodic signal of any form [9]. For the function $f$ defined in the interval $[-1, 1]$, the trigonometric series has the form

$$f(x) = \frac{a_0}{2} + \sum_{n=1}^{\infty} \left( a_n \cos\frac{\pi n x}{l} + b_n \sin\frac{\pi n x}{l} \right),$$

where

$$a_n = \frac{1}{l}\int_{-l}^{l} f(x)\cos\frac{\pi n x}{l}\,dx, \ n = 0,1,2,...,$$

$$b_n = \frac{1}{l}\int_{-l}^{l} f(x)\sin\frac{\pi n x}{l}\,dx, \ n = 0, 1, 2,...,$$

Solvation shell deformation and therefore the formation of the aquacomplex of the polarization charge create the conditions for the action of the constant component of the field, and therefore, for the excitation of the oriented drift of the aquacomplex [10, 11].

It should be assumed that the field frequency should correlate with the natural frequency of the system $[Me(OH_2)_n]^{m+}$, where Me is the metal ion, $m$ is the multiplicity of its charge, and $n$ is the coordination number. The energy of the aquatic complex, which characterizes its state as a system, consists of four parts:

• translational energy of the entire aquacomplex;

• electronic energy determined by the quantum state of the electron shells of the cation and $OH_2$ solvate groups;

• vibrational energy due to oscillation of the cation and solvate groups relative to each other;

• rotational energy corresponding to the rotation of the entire aquacomplex or its parts relative to each other.

In accordance with the foregoing, the natural frequency of aquacomplex (solvated cation) can be expressed by the following relation: $v = v_e + v_c + v_s$, in which the terms correspond excitation frequencies of electron shells ($e$), oscillations of the cation and solvate groups ($c$) and rotation of solvate groups relative to the cation ($s$). It is natural to assume that the excitation of electron shells or oscillations of the cation and solvate groups can be initiated only after deformation of the solvation shell (screen), preventing the action of an electric field. The deformation, essentially, to the rotation of the solvate groups relative to the cation, corresponds to the term $v_s$.

Polarization and, consequently, deformation of the solvation shell leads to the formation of a system similar to a system of two oppositely charged particles of different masses. For example, the polarized cationic aquacomplex $[Li(OH_2)_4]^+$ can form two systems: $^6_3Li^+-4 (OH_2)$ and $^7_3Li^+-4(OH_2)$. They differ in the masses of the positive part and, therefore, the location point of the centre of inertia of the dipole. When an electric field is applied to these systems, the vibrational and rotational components of motion are excited. During the first half-cycle, the angular momentum increases, during the second half it is partially compensated. The uncompensated part of the moment is transformed into the translational component of the motion of the centre of inertia. The speeds of translational motion (drift) of isotopically distinct complexes in the siluric of different inertial properties will also differ.

## 2.1. Model of the effect of electric and magnetically induced selective drifts

The law of conservation of momentum allows one to naturally formulate the concepts of rest and speed of a mechanical system as a whole. The statement about the additivity of mass says that the connection between the momentum **P** and the speed **V** of the system as a whole is the same as between the momentum and the speed of one material point by the mass equal to the sum of the masses of all particles in the system [12].

The speed of the system as a whole is the speed of movement of the centre of inertia of the system. In addition, during the movement of a closed system, the angular momentum of the system is preserved:

$$\mathbf{M} = \sum_i [\mathbf{r}_i \mathbf{p}_i] = \text{const},\tag{2.1}$$

moreover, like in a pulse, it does not depend on the presence or absence of interaction between the particles.

In the mathematical formulation of the process model, the polarized cationic aquacomplex is 'placed' in the $XOY$ plane. The vectors of forces acting on the positive and negative parts of the polarized aquacomplex (dipole) are thus located in the same plane and have a single nonzero component. They are parallel to the field strength vector, i.e., axis $OX$, and have opposite directions. This is represented schematically in Fig. 2.1.

The electric field vector is also located in the $XOY$ plane and has a nonzero component $E_x$. In perpendicular planes: the first (grounded) is located in $YOZ$, and the second (potential) electrode is in the $YOZ$ plane. Component $E_x$ – alternating, which is a function of time (Fig. 2.2).

The main idea of a numerical description of the selective drift process is to split the initial non-stationary system of equations into physical processes [13]. The whole process of computing consists of multiple repetition of time steps. The calculation of each time step $2\Delta t$ is divided into three stages [14].

At the first stage (half-step $\Delta t$), the increment of the angular momentum due to the action of the first half-cycle of the electric field $\Delta M_1$ is considered. The projections of the pulses of the positive and negative parts of the aquacomplex at the first stage are determined by the relations

$$p_+ = E^+ q_+ \Delta t,$$
$$p_- = -E^+ q_- \Delta t.\tag{2.3}$$

and the negative parts of the aquacomplex in the second stage are

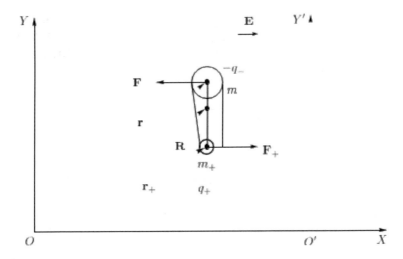

**Fig. 2.1.** Polarized aquacomplex in an electric field: $q$ and $m$ are the absolute values of the charge and mass of the negative (–) and positive (+) parts, respectively

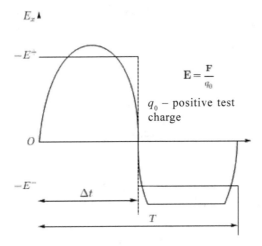

**Fig. 2.2.** Asymmetric electric field intensity: $E^+$ and $E_-$ are the effective values of the intensity in the positive and negative half-periods respectively.

determined by the relations

$$p_+ = -E^- q_+ \Delta t,$$
$$p_- = -E^-(-q_-) \Delta t = E^- q_- \Delta t. \qquad (2.3)$$

In the third stage, the uncompensated part of the moment is determined $\Delta M = \Delta M_1 + \Delta M_2$, translational speed $V$ and, therefore, obviously, the displacement of the centre of inertia.

Figure 2.3 shows the proposed type of trajectories of isotopically different cationic aquacomplexes located in an asymmetric electric

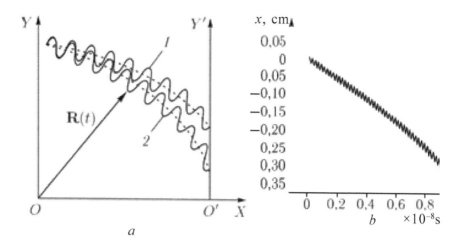

**Fig. 2.3.** *a* – the proposed type of trajectories $R(t)$ of isotopically distinct aquacomplexes: the figure illustrates the process at a qualitative level; *b* – calculated dependence of the coordinate of the centre of mass of the diatomic dipole.

field. The mass of the positive part of the aquacomplex 2 is greater than in aquacomplex 1.

Cationic aquacomplex 1 drifts in the *OX* direction faster and earlier reaches the plane *YOZ*. Aquacomplex 2 drifts more slowly. The full path of the heavier aquacomplex 2 is longer. In addition, the path forming the trajectories substantially deviates from the first one. The enrichment by the aquacomplexes of the first type is observed over the *YOZ* plane.

.In fact, given the chaotic thermal motion,the generators of the trajectories (arcs) will be broken, and the directions of the displacements will stochastically change in the *YOZ* and *YOX* planes.The ratio of the displacements due to the field-induced rotational-translational motion and displacement values caused by chaotic motion is determined by many parameters. Of them we can distinguish the field strength *E* and the temperature of the solution *T*.

With an increase in the *E/T* ratio, a decrease in the stochastic component of the movement should be expected. In addition, even in the hypothetical ideal case (the absence of thermal motion) the form of the trajectories can differ significantly from that shown in Fig. 2.3 *a*. It is determined by a set of (at least) 7 parameters:

- field strength;
- frequency;
- asymmetry coefficient;

• communication length in a dipole (polarized aquacomplex);
• mass ratio of the positive and negative parts of the dipole;
• the mass of the aquacomplex as a whole;
• polarization coefficient of the aquacomplex.

Figure 2.3 *b* illustrates the above. It shows the time dependence of the shift of the centre of mass of the $^6Li-^1H$ dipole. It is one of the simplest cases for a free diatomic molecule in an asymmetric electric field with parameters: $E^+ = 700$ V/m; $E^- = 1000$ V/m; $v = 500$ kHz. The dipole polarization coefficient in the calculations was taken equal to 0.067. It can be seen that the centre of mass of the dipole in this case moves towards a grounded electrode. Every time at a change in the sign of the field strength there occurs reorientation of the dipole – rotation, which determines the nature of the time dependence of the centre of the displacement of the mass

It is clear that such a combination of parameters should be provided when during one half-period of the field the aquacomplex 'did not keep up' to make a complete rotation about the centre of inertia. Also it is seen that with certain combinations of field and solution parameters we should expect a 'disappearance' of the drift effect: the aquacomplex will be rotated relative to the centre of inertia at almost zero translational component of the movement.

The speeds of translational motion (drift) of different aquacomplexes in the direction $OX$ because o different inertial properties will differ. Figure 2.4 schematically shows the sequence (1–8) of changes in the structure of a solution under the action of an asymmetric electric field, which is formed by two plane electrodes between which the solution is placed. The letters '*A*' and '*B*' denote areas in the volume of the solution within which sampling is performed. At the initial time (1), the concentration of solvated ions of two different types (they differ in colour) in the whole solution and within the sampling areas are equal. The action of an external field on the solution excites mass transfer in the volume of the solution. Differences in the inertial properties of the solvated ions lead to the fact that the structure of the solution becomes heterogeneous, and in its volume regions with a predominant content of one of the two types of ions are formed. In the end, after some time (7) and (8), the solvated ions of exclusively one type are located in each of the sampling regions. The solution becomes enriched with one of two types of ions in a certain spatial region. Conditions are

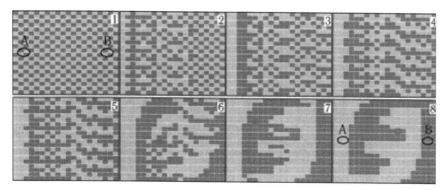

**Fig. 2.4.** Excitation of mass transfer in volume and change in the structure of a solution of a mixture of two salts.

created for the selection of a mixture of two salts from the solution volume of only those ions that are needed. To do this, it is necessary to select from a 'suitable' point in the volume of the solution.

It should be expected that the frequency of the field at which the transformation of the rotational moment into translational motion is maximum should correlate with the natural frequency of the system [I $(OH_2)_n]^m$, where I is the ion (cation or anion), $m$ is the multiplicity of its charge, $n$ is the number of water molecules in the solvation shell.

The total energy of the aquacomplex, which characterizes its state as a system, is characterised by four components:

• energy of translational motion of the entire aquacomplex;

• electron energy, determined by the quantum state of the electron shells of the ion and solvate groups $(OH_2)$;

• vibrational energy due to ion oscillation and the solvate groups relative to each other;

• rotational energy corresponding to the rotation of the entire aquacomplex or its parts relative to each other.

At the same time, the inertial properties of aquacomplexes may vary for several reasons.

1) Due to the difference in the mass of cations with equal coordination numbers. An example would be aquacomplexes formed by cations of lithium isotopes $[Li^6(OH_2)_4]^+$ and $[Li^7(OH_2)_4]^+$, or aquacomplexes formed by cations of calcium and magnesium $[Ca^{40}(OH_2)_4]^{2+}$ and $[Mg^{24}(OH_2)_4]^{2+}$. In the case of lithium, the effect can be used for enrichment on the necessary isotope, in the case of calcium–magnesium – for the separation of cations of calcium and magnesium.

2) Due to the difference in the masses of anions with equal coordination numbers. Example: $SO_4^{2-}$ ion forms isotopically excellent aquacomplexes $[S^{32}O_4(OH_2)_6]^{2-}$ and $[S^{34}O_4(OH_2)_6]^{2-}$, and the dissolution of the mixture of nitrate and periodate leads to the formation of a mixture of aquacomplexes $[NO_3(OH_2)_4]^-$ and $[IO_4(OH_2)^4]^-$. In the case of sulphur, the effect may be used to separate the isotopes $S^{32}$ and $S^{34}$, and in the case of nitrogen–iodine for selective cleaning of the solution from nitrate or periodate.

3) Due to the difference in the coordination numbers of ions, which also can be used for elemental enrichment of the solution.

In the mathematical description of the process of excitation of mass transfer in a salt solution under the action of a periodic magnetic field, the amplitude of the cluster displacement under the influence of a magnetic field can be estimated using the following algorithm [15]:

1) the modulus of the cluster velocity vector corresponding to the thermal (Brownian) motion in the solution is determined;

2) the 'deformation' uncompensated cluster charge caused by the deformation of the solvation shell in collisions is estimated;

3) under the assumption that the average velocity vector of the clusters is directed at a certain angle (nonzero) to the magnetic induction vector (magnetic field strength), the absolute value of the Lorentz force acting on the cluster between the Brownian collisions is determined;

4) the frequency of Brownian collisions is determined and, therefore, the time between collisions;

5) the momentum acquired due to the action of the magnetic field by the deformed cluster between the Brownian collisions is determined, and then the cluster is displaced relative to the segment of its trajectory between the collisions.

The amplitude of such a shift is primarily determined by the strength of the magnetic field and the ratio of the frequencies of Brownian collisions and the alternating magnetic field. If the amplitude of the magnetically induced bias exceeds the average range of the clusters in the solution, then directional mass transfer is possible.

Since the clusters are formed by both cations and anions, the uncompensated charge caused by the deformation of the solvation shell can be both positive and negative. A diode is required to ensure the possibility of 'exchange' of excess charges and the closure of the current circuit in the 'turn' of the solution.

## 2.2. Cluster structure of the solution and excitation frequencies of electroinduced selective drift

The structure of the solution can be represented as an ensemble of ordered supramolecular formations – clusters (they will be called structural units), which constantly lose some of their constituent parts and are replenished by others, can be destroyed and created again. The fluid structure is statistical in nature. These formations as a whole can move in volume, make oscillatory and rotational movements. In the general case, the structural units of a liquid can be represented as systems of a large canonical ensemble and include $g$ monomers.

The solvent molecules that make up the monomer interact with each other through several channels, which are listed below.

**Intermolecular interaction** – the interaction between electrically neutral molecules or atoms. The forces of intermolecular interaction were first taken into account by J.D. van der Waals to explain the properties of real gases and liquids. Intermolecular interaction is of an electrical nature and consists of attractive forces (orientational, induction and dispersive) and repulsive forces.

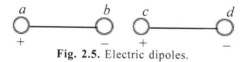

**Fig. 2.5.** Electric dipoles.

**Orientational forces.** Two electric dipoles $ab$ and $cd$ at the indicated mutual arrangement are attracted, since opposite charges at points $b$ and $c$ interact more strongly than like charges at points $a$ and $c$ (as well as at $b$ and $d$). The orientational forces act between polar molecules, i.e., possessing dipole electrical moments. The force of attraction between between two polar molecules is maximized when their dipole moments are located along one line (see Fig. 2.5). This force arises due to the fact that the distances between the opposite charges are slightly less than those between the opposite ones. As a result, the attraction of dipoles exceeds their repulsion. The interaction of dipoles depends on their relative orientation, and therefore the dipole interactions are called *orientational*. The chaotic thermal motion continuously changes the orientation of the polar molecules, but, as the calculation shows, the average value of the force over all possible orientations has a certain value that is not equal to zero. The potential energy of the orientational intermolecular interaction is determined by the relation

$$U_{or}(r) \sim \frac{p_1 p_2}{r^6},$$

where $p_1$, $p_2$ are the dipole moments of interacting molecules. Accordingly, the force of orientational interaction $F_{or} \sim r^{-7}$. The $F_{or}$ force decreases with distance much faster than the Coulomb interaction force of charged bodies ($F_q \sim r^{-2}$).

**Induction forces.** Induction (or polarization) systems act between the polar and nonpolar molecules. A polar molecule creates an electric field that polarizes the molecule with electric charges uniformly distributed throughout its volume. Positive charges are displaced in the direction of the electric field (that is, from the positive pole), and negative charges are opposed to the positive pole. As a result, a dipole moment is induced in a nonpolar molecule.

The intermolecular interaction energy in this case is proportional to the dipole moment $p_1$ of the polar molecule and the polarizability $a_2$, which characterizes the ability of the other molecule to polarize:

$$U_{ind}(r) \sim \frac{p_1 a_2}{r^6}.$$

This energy is called *induction*, as it appears due to the polarization of molecules caused by electrostatic induction. Induction forces ($F_{in} \sim r^{-7}$) also act between polar molecules.

**Dispersion forces.** Between nonpolar molecules there is a dispersion intermolecular interaction. The nature of this interaction can only be partially explained using **quantum mechanics**. In atoms and molecules, **electrons** move in complex ways around nuclei. As a result (the average time) the dipole moments of nonpolar molecules turn out to be equal to zero. But at every fixed point in time, the electrons occupy some position. Therefore, the instantaneous value of the dipole moment (for example, for a hydrogen atom) is nonzero. An instant dipole creates an electric field that polarizes neighbouring molecules. As a result, the interaction of 'instant dipoles' arises. The interaction energy between nonpolar molecules is the average result of the interaction of all kinds of 'instantaneous dipoles' with the dipole moments that they induce in the neighbouring molecules due to induction. The potential energy of dispersion intermolecular interaction is

$$U_{diso}(r) \sim \frac{a_1 a_2}{r^6},$$

where $a_1$, $a_2$ are the polarizabilities of the interacting molecules. Dispersion interaction force $F_{disp} \sim r^{-7}$.

The intermolecular interaction of this type is called *dispersive* because the dispersion of light in a substance is determined by the same properties of the molecules as this interaction. The dispersion forces act between all atoms and molecules, since the mechanism of their appearance does not depend on whether at the molecules (atoms) there are constant dipole moments or not. Typically, these forces are superior in magnitude to both orientational and induction ones. Only in the interaction of molecules with large dipole moments, for example, water molecules, $F_{or} > F_{disp}$ (3 times for water molecules). In the interaction of such polar molecules as CO, HI, HBr and others, dispersion forces are tens and hundreds of times superior to all others.

It is very important that all three types of intermolecular interaction decrease in the same way with distance:

$$U = U_{or} + U_{ind} + U_{disp} \sim r^{-6}.$$

**Repulsive forces**. The repulsive forces act between molecules at very short distances, when the filled electronic shells of the atoms that make up the molecules come into contact. The Pauli principle existing in quantum mechanics prohibits the penetration of filled electron shells into each other. The repulsive forces arising from this depend to a larger degree than the attractive forces, on the individuality of the molecules. A good agreement with the experimental data is obtained assuming that the potential energy of the repulsive forces increases with decreasing distance according to the law

$$U_{rep}(r) \sim r^{-12},$$

and the actual value of the force $F_{rep} \sim r^{-13}$.

**The resulting interaction.** It is almost impossible to calculate $U(r)$ with sufficient accuracy using the apparatus of quantum mechanics with a huge variety of pairs of interacting molecules. It is still not possible to experimentally measure the strength of the interaction at intermolecular distances. Therefore, they usually choose a formula for $U(r)$ such that the calculations would agree well with the experiment. The most commonly used dependence of the potential $U(r)$ of intermolecular interaction on the distance $r$ between the intermolecules:

$$U(r) \simeq 4\varepsilon \left[ \left( \frac{\sigma}{r} \right)^{12} - \left( \frac{\sigma}{r} \right)^{6} \right],$$

with the so-called *Lennard–Jones* potential. In this formula, the distance $r = \sigma$ is the smallest possible distance between the stationary molecules, $\varepsilon$ is the depth of the 'potential well' (the binding energy of the molecules). The values of $\sigma$ and $\varepsilon$ included in the formula are determined experimentally based on the dependence of the properties of substances (for example, diffusion, thermal conductivity, or viscosity).

**Exchange interaction.** This is a specific mutual influence of identical particles, which is effectively manifested as a result of some special interaction. The exchange interaction is a purely quantum-mechanical effect that has no analogue in classical physics. Due to the quantum-mechanical principle of indistinguishability of identical particles, the wave function of the system must have a certain symmetry with respect to the permutation of two identical particles, that is, their spatial coordinates and spins: for particles with an integer spin – bosons – the wave function of the system does not change under such a permutation (it is symmetric), and for particles with a half-integer spin – fermions – it changes sign (is antisymmetric). It should be remembered that the same particle, for example an electron, can obey both the Fermi statistics (fermion) and the classical Boltzmann statistics. It depends on which part of the system the electron in question is. An electron that is part of an atom (a bound electron) or electrons in a solid obey Fermi statistics and are fermions. Free electrons, for example, electrons of a rarefied hot plasma, obey Boltzmann statistics. If the interaction forces between the particles do not depend on their spins, the wave function of the system can be represented as the product of two functions, one of which depends only on the coordinates of the particles, and the other only on their spins. In this case, from the principle of identity it follows that the coordinate part of the wave function, which describes the motion of particles in space, must have a certain symmetry relative to a permutation of the coordinates of identical particles, depending on the symmetry of the spin function. The presence of such symmetry means that there is a certain consistency (correlation) of motion of identical particles, which affects the energy of the system (even in the absence of any force interactions between particles). Since usually the influence of particles on each other is the result of the action of some forces between the two, the mutual influence of identical particles resulting from the principle of identity is said to be a manifestation of a specific interaction – exchange interaction.

If we consider a liquid as a system of electrically non-interacting neutral identical molecules, each of which has its own, undeformed, generalized electronic shell, then the molecules will obey Boltzmann statistics. If we take into account the fact that the generalized electronic shell of a molecule can be deformed for one reason or another, which causes the formation of a polarization charge, then it is necessary to use either the Fermi statistics or the Bose statistics.

Electrons, positrons, protons and neutrons are fermions. An example of a boson is a photon. In the general case, a particle consisting of an odd number of fermions is also a fermion, and a particle consisting of an even number of fermions is a boson. Fermions have a half-integer spin, and bosons a whole spin.

We can try to illustrate the appearance of the exchange interaction by the example of two molecules with a polarization charge '$e$' and obeying Fermi statistics. We assume that their spin interaction is small; therefore, the wave function $\Psi$ of two molecules can be represented as

$$\Psi = \Phi(r_1, r_2) \, \chi(s_1, s_2), \tag{2.4}$$

where $\Phi(r_1, r_2)$ is a function of the coordinates $r_1$, $r_2$ of the molecules, and $\chi(s_1, s_2)$ are functions of the projection of their spins $s_1$, $s_2$ in a certain direction. Since polarized molecules are regarded as fermions, the total wave function $\psi$ must be antisymmetric. If the total spin of both molecules is zero (the spins are antiparallel), then the spin function is antisymmetric with respect to the permutation of the spin variables and, therefore, the coordinate function $\Phi$ must be symmetric with respect to the permutation of the coordinates of the molecules.

If the total spin of the system is 1 (the spins are parallel), then the spin function is symmetric, and the coordinate one is antisymmetric. Denoting by $\psi_n(r_1)$ and $\psi_n(r_2)$ the wave functions of individual molecules (indices $n$, $n'$ mean a set of quantum numbers that determine the state of polarized molecules in the ion field), we can neglect first the interaction between the generalised electron shells of the molecules and write the coordinate part of the wave function in the form

$$\Phi_a = \frac{1}{\sqrt{2}}\left[\psi_n(r_1)\psi_{n'}(r_2) - \psi_{n'}(r_1)\psi_n(r_2)\right] \quad \text{for the case with } S = 1,$$

$$\Phi_S = \frac{1}{\sqrt{2}}\left[\psi_n(r_1)\psi_{n'}(r_2) + \psi_{n'}(r_1)\psi_n(r_2)\right] \quad \text{for the case with } S = 0.$$

(multiplier $1\sqrt{2}$ introduced for the normalisation of the wave function). In the state with the antisymmetric coordinate function $\Phi_a$, the average distance between the generalized electron shells is greater than in the state with the symmetric function $\Phi_S$; this is clear from the fact that the probability $|\Psi|^2 = |\Phi_a|^2|\chi_S|^2$ of the location of polarized molecules at the same point $r_1 = r_2$ for the state $\Phi_a$ is equal to zero. Therefore, the average energy of the Coulomb interaction (repulsion) of two molecules turns out to be lower in the state $\Phi_a$ than in the state $\Phi_S$. The correction to the energy of the system associated with the interaction of electron shells can be determined by perturbation theory and is equal to

$$E_{\text{int}} = K \pm A, \tag{2.6}$$

where the signs $\pm$ refer respectively to the symmetric $\Phi_S$ and antisymmetric $\Phi_a$ coordinate states,

$$K = e^2 \int \frac{|\psi_n(r_1)|^2 |\psi_{n'}(r_2)|^2}{|r_1 - r_2|} dV_1 dV_2,$$

$$A = e^2 \int \frac{|\psi_n^*(r_1)|^2 \psi_{n'}(r_1)\psi_n^*(r_2)\psi_n(r_2)}{|r_1 - r_2|} dV_1 dV_2,$$

$$\tag{2.7}$$

($dV = dx\,dy\,dz$ is the volume element). The value of $K$ has a clear visual classical meaning and corresponds to the electrostatic interaction of two charged 'clouds' with charge densities $e|\psi_n(r_1)|^2$ and $e|\psi_n(r_2)|^2$. The quantity $A$, called the exchange integral, can be interpreted as the electrostatic interaction of charged 'clouds' with charge densities $e\psi_n^*(r_1)\psi_{n'}'(r_1)$ and $e\psi_n'(r_1)\psi_n'(r_2)\psi_n(r_2)$, that is, when each of the molecules is simultaneously in states $\psi_n$ and $\psi_n'$ (which is pointless from the point of view of classical physics). It follows from (2.6) that the total energy of the system of two molecules can differ by $2A$. Thus, although the direct spin interaction is small and not taken into account, the identity of the two molecules leads to the fact that the energy of the system turns out to be dependent on the full spin of the system, as if between the molecules there was an additional (exchange) interaction. Obviously, the exchange interaction in this case is part of the Coulomb interaction of generalized electron shells and explicitly it appears even in an approximate consideration of a quantum-mechanical system, when the wave function of the entire system is expressed in terms of the wave functions of individual particles.

The exchange interaction is effectively manifested when the wave functions of the individual particles of the system 'overlap', that is, when there are regions of space in which the particle can be with a noticeable probability in different states of motion. This can be seen from the expression for the exchange integral $A$: if the degree of overlap of the states $\psi_n^*(r)$ and $\psi'_n(r)$ is negligible, then the quantity $A$ is very small.

From the principle of identity it follows that the exchange interaction arises in a system of identical particles even if the direct force interactions of the particles can be neglected, that is, in an ideal gas of identical particles. Effectively, it begins to manifest itself when the average distance between the two particles becomes comparable (or shorter) than the de Broglie wavelength corresponding to the average particle velocity. The nature of the exchange interaction is different for fermions and for bosons. For fermions, the exchange interaction is a consequence of the Pauli principle, which prevents the convergence of identical particles with the same direction of spins, and is effectively manifested as repelling them from each other at distances of the order of or less than the de Broglie wavelength. The nonzero energy of the degenerate fermion gas (Fermi gas) is entirely due to such an exchange interaction. In a system of identical bosons, the exchange interaction, on the contrary, has the character of mutual attraction of particles. In these cases, consideration of systems consisting of a large number of identical particles is based on quantum Fermi–Dirac statistics for fermions and Bose–Einstein statistics for bosons.

If the interacting identical particles are in an external field, for example, the solvent molecules in the Coulomb field of an ion, then the existence of a certain symmetry of the wave function and, accordingly, a certain correlation of the movement of particles affects their energy in this field, which is also an exchange effect. This exchange interaction should make a contribution of the opposite sign in comparison with the contribution of the exchange interaction of particles with each other. Therefore, the total exchange effect can both lower and increase the total interaction energy in the system. The energy profitability or the disadvantage of a state with parallel fermion spins depends on the relative values of these contributions. For example, in moleculs with a covalent chemical bond, for example, in an $H_2$ molecule, a state in which the spins of the valence electrons of the connecting atoms are antiparallel is energetically favourable.

With fluctuations in a given volume in each monomer – cluster dispersion $(\Delta g)^2$ is determined by the average number of molecules $(g_a)$: $(\Delta g)^2 = g_a$, and volume fluctuations $((\Delta v)^2_n)^{1/2}$ and the number of molecules $(g^{1/2})$ and the relative volume fluctuation $(\delta_V)$ are related by $\delta V = (\Delta v)^2_n)^{1/2}/v = 1/g^{1/2}$.

The general motion of the cluster can be represented as the translational movement of its centre of inertia and the movement of components (fragments) relative to this centre. The latter in the presence of an external field can be both oscillatory and rotational. In this case, the total energy $(E)$ of the cluster can be represented as the sum of the kinetic $E_{kin} = (1/2)I(d\theta/dt)^2$ and potential $U = 2U_0\sin^2(\theta/2)$ energy, where $I$ is the moment of inertia, $U_0$ is the amplitude of the potential energy component, $\theta$ is the angle of rotation of the cluster relative to the field direction. In the case when the total energy does not exceed $2U_0$, the cluster performs oscillatory movements that can go (convert) into rotational ones when the condition $E \geq 2U_0$ is reached, which is possible due to fluctuations like the potential barrier, as well as cluster energy under the influence of external electromagnetic and thermal fields. The frequency of vibrational movements is determined by the equation $v_0 = (1/(2\pi))(U/I)^{1/2}$, multiplying both sides of which by $h$, assuming that $hv_0 = E_0$, $U0 \cong E/2$, $E_0 \cong E$ we obtain the formula for the frequencies of transition of vibrational movements into rotational

$$v0 \cong h/(8\pi^2 I). \tag{2.8}$$

The latter in conditions of achieving relative mobility by the clusters, necessary for activating the selective drift process under the influence of a field, can be considered as the lowest frequencies of the cluster free motions. A wider range of frequencies of natural motions of clusters associated with free rotational–translational movements, can be represented as a set of frequencies of allowed quantum levels of free rotation [16]

$$v = J(J + 1)h/(8\pi^2 I) \qquad (2.9)$$

and the translational movement of the body

$$v = (n_x^2 + n_y^2 + n_z^2)h/(8\mu a^2), \qquad (2.10)$$

where $J$ is the quantum number, $n_{x,\,y,\,z}$ are the projections of the rotation vector on the coordinate axes, $a$ is the cluster size, $\mu$ is the reduced mass of the cluster.

If in formulas (2.9) and (2.10) the cluster shape is assumed to be spherical (the 'zero' approximation), then we can assume that $J = 1$, $(n_x^2 + n_y^2 + n_z^2) = 1$. The number of water molecules (solvate groups) in the cluster, having a radius $r_{cl}$, is $g$, $a = 2r_{cl}$, $r_{cl} = rg^{1/3}$, $r = (3V/(4\pi))^{1/3}$, where $r$, $V$ is the radius and volume of the solvate group, respectively; $I = K_r \mu r_{cl}^2$, $K_r$ is a coefficient having a value of 0.4 during rotation of the solvation shell around the cation and 1.4 when rotating about an axis passing along the surface of the cluster.

According to the general principles of quantum mechanics for a linear system of rigidly bound particles, the frequency corresponding to the fundamental rotational state, determined by the relation [17]

$$v_F = \frac{h}{4 \cdot \pi^2 \cdot I}, \qquad (2.11)$$

where $I = \dfrac{m_1 \cdot m_2}{m_1 + m_2} \cdot r_0^2$ is the moment of inertia of this system; $m_1$ and $m_2$ are the masses of the particles; $r_0$ is the distance between them; $h$ is Planck's constant.

In the very first approximation, to determine the order of magnitude of $v_p$, as a calculated analogue for the solvated cation, we will use a two-particle system with similar properties. One of the particles of the system is a cation, the second is a solvation shell, formed by the solvate groups. Even with such a rough looking there are three complex questions regarding the structure of the aquacomplex.

First: how far is the distance between the cation and the solvation shell?

If we assume that it is equal to the radius of the solvation shell, then the following question arises – which solvation shell: due to primary or due to secondary hydration?

Second: how many are the number of solvate groups in the solvate shell, that is, how much is its mass? If we assume that the field acts on the solvation shell formed by primary hydration, then the number of solvation groups should correspond to the coordination number of the cation. If the field acts on the shell formed by the secondary hydration, then additional assumptions are required.

Third question: is it possible to confine ourselves to the consideration of secondary hydration? To date, the mechanism of ion–molecular interaction remains unclear. According to established beliefs in electrochemistry 'the size of the ionic shells in aqueous electrolytes does not exceed several tens of water molecules' [18]. Further, the ion–dipole interaction energy rapidly weakens and cannot provide for the structuring of polar water molecules in an electric field of the central cation. In the same work, it is indicated that 'the neutralizing effect is exerted by thermal motion and counterions forming the ionic atmosphere'. Despite this, the main postulate of the electrodynamics of weakly conducting liquids is the postulate of the 'freezing-in' of a space charge. It is confirmed by a number of simple experiments described in [19], they indicate the presence of 'hard' bonds of ions and neutral molecules in weakly conducting media. Thus, the radius of the aquacomplex is at least ten sizes of water molecules. Within a sphere with such a radius there will be at least $10^4$ solvate groups. So, if the radius of the aquacomplex is 100 diameters of the water molecule (about $1.93 \cdot 10^{-6}$ cm), then the volume of the sphere formed by such a radius should be about $9 \cdot 10^5$ solvate groups. There are also experimental data that indicate that the introduction of a bulk electric charge into a weakly conducting liquid is accompanied by the formation of supramolecular formations. From the estimates of the attached mass water, structured around each ion, it follows that the linear dimensions of such a supramolecular formation – 'cluster' – are about 1 µm. In a sphere with a volume equal to the volume of this cluster, there should be more than $10^{10}$ water molecules.

Analysis can be carried out by the aquacomplexes formed by cations of high values of the coordination number. The greater $n$, the greater the mass of the solvate shell, the moment of inertia of the solvated cation, and, therefore, the smaller the component of the natural frequency of the analyzed system, which corresponds to the rotation of the solvate groups relative to the cation. Relatively low frequencies cause far fewer problems, both when setting up an experiment and in the case of implementing a process on a technologically significant scale.

In this regard, elements of the third group of the periodic system, or rather, its side subgroup: Sc, Y, La, and Ac are of interest. Their coordination number can take values from 6 to 12 [20]. This, on the one hand, is convenient, and on the other hand, solvated cations having different inertial properties are present in the same salt solution of one metal of the third group. For example, in the aqueous solution of lanthanum nitrate there are $[La(OH_2)_6]^{3+}$ and $[La(OH_2)_9]^{3+}$ aquacomplexes. The only thing known is that there are more $[La(OH_2)_9]^{3+}$ aquacomplexes – they are most represented. The probability of formation of aquacomplexes for coordination numbers 4, 8, and 12 is very small, but not excluded.

To determine the value of $v_p$, it is necessary to choose the approximation in accordance with which the value $r_0$ will be set – the distance between the cation and the solvate shell associated with a single quasiparticle. Suppose that $r_0$ is equal to the distance at which the electric ternary field of the cation does not cause polarization of the water molecules.

It is known that the value of hydration energy through the Born equation [20] is associated with the value of the radius of the sphere within which the water molecules are structured:

$$-\Delta G_H = \frac{e^2 \cdot z^2}{2 \cdot r_0} \left(1 - \frac{1}{\varepsilon}\right),$$

where $e$ is the electron charge, $z$ is the multiplicity of the ion (cation) charge, $\varepsilon$ is the dielectric constant of water. Table 2.1 shows the $v_p$ values obtained in the above approximations for various ions. The values of the excitation frequencies of the rotational component of the movements of aquacomplex parts are obtained for $r_0$ corresponding to the energy of primary $\Delta G_H^I$ and secondary hydration $\Delta G_H^{II}$. Here it was assumed that the number of solvate groups in the shell of secondary hydration is $2n$. As can be seen from the table, the values of the excitation frequencies depend both on the mass of the hydration shell (the number of attached water molecules – solvate groups) and on the distance between the cation and the solvate shell.

The ionic radius is uniquely associated with the value of the coordination number, which is implemented in the aquacomplex. In particular, the ionic radius for $n = 12$ is 1.12 times greater than that for $n = 6$; for $n = 8$ it is 1.03 times, and for $n = 4$ it becomes smaller: $r_0^{(4)} = 0.94 \cdot r_0^{(6)}$. The estimates show that for the same coordination number, an increase in the cation mass by 55% leads to a decrease the frequency of excitation of the rotational motion of the solvation shell relative to the cation by 40%.

**Table 2.1**. Frequency characteristics of rotational bond excitation: cation – solvation shell; $v_p$, Hz

| Ion | Ion radius Å (for $n = 6$) | | Binding frequency Me-OH$_2$ | Me-(OH$_2$)$_n$ binding frequency for primary hydration | Me-(OH$_2$)$_n$ binding frequency for secondary hydration |
|---|---|---|---|---|---|
| Li$^+$ | 0.68 | 4 | $4.896 \cdot 10^{10}$ | $3.874 \cdot 10^{10}$ | $1.309 \cdot 10^{10}$ |
| Na$^+$ | 0.97 | 6 | $1.692 \cdot 10^{10}$ | $9.014 \cdot 10^{9}$ | $3.346 \cdot 10^{9}$ |
| K$^+$ | 1.33 | 6 | $1.043 \cdot 10^{10}$ | $4.478 \cdot 10^{9}$ | $1.728 \cdot 10^{9}$ |
| Rb$^+$ | 1.47 | 8 | $7.893 \cdot 10^{9}$ | $2.188 \cdot 10^{9}$ | $8.239 \cdot 10^{8}$ |
| Cs$^+$ | 1.67 | 8 | $6.568 \cdot 10^{9}$ | $1.507 \cdot 10^{9}$ | $5.359 \cdot 10^{8}$ |
| Be2+ | 0.35 | 4 | $6.350 \cdot 10^{10}$ | $4.762 \cdot 10^{10}$ | $1.293 \cdot 10^{10}$ |
| Mg$^{2+}$ | 0.66 | 6 | $2.325 \cdot 10^{10}$ | $1.212 \cdot 10^{10}$ | $3.883 \cdot 10^{9}$ |
| Ca$^{2+}$ | 0.99 | 6 | $1.376 \cdot 10^{10}$ | $5.845 \cdot 10^{9}$ | $2.033 \cdot 10^{9}$ |
| Sr$^{2+}$ | 1.12 | 8 | $9.913 \cdot 10^{9}$ | $2.717 \cdot 10^{9}$ | $9.295 \cdot 10^{8}$ |
| Ba$^{2+}$ | 1.34 | 8 | $8.375 \cdot 10^{9}$ | $1.896 \cdot 10^{9}$ | $6.310 \cdot 10^{8}$ |
| Al$^{3+}$ | 0.51 | 6 | $2.622 \cdot 10^{10}$ | $1.312 \cdot 10^{10}$ | $3.885 \cdot 10^{9}$ |
| Y$^{3+}$ | 1.06 | 9 | $1.123 \cdot 10^{10}$ | $2.928 \cdot 10^{9}$ | $9.813 \cdot 10^{8}$ |
| La3+ | 1.14 | 9 | $8.364 \cdot 10^{9}$ | $1.782 \cdot 10^{9}$ | $6.018 \cdot 10^{8}$ |

The $r_0$ value determined for the solvate shell during the primary hydration of the La$^{3+}$ cation is 2.74 Å for $n = 9$. If we assume that the rotational movement of the solvate groups OH$_2$, which are associated in the supramolecular formation (in the cluster), is excited, then at least ten diameters of the water molecule should be taken as the value of $r_0$, and the mass of structured water in the shell should be considered equal to about 100 masses of water molecules. Estimates show that, under these assumptions, the $v_p$ value corresponds to about

**Table 2.2.** Hypothetical frequencies of excitation of the rotational movement of solvate groups in the aquacomplex; $v_p$, Hz

| Ion | $r_0 = 50d_{H2O}$ $m_2 = 10^3 m_{H2O}$ | $r_0 = 50d_{H2O}$ $m_2 = 10^5 m_{H2O}$ | $r_0 = 100d_{H2O}$ $m_2 = 10^3 m_{H2O}$ | $r_0 = 100d_{H2O}$ $m_2 = 10^5 m_{H2O}$ | $r_0 = 0.5\ \mu m$ (cluster) $m_2 = 10^3 m_{H2O}$ | $r_0 = 0.5\ \mu m$ (cluster) $m_2 = 10^5 m_{H2O}$ |
|---|---|---|---|---|---|---|
| $Li^+$ | $1.549 \cdot 10^7$ | $1.548 \cdot 10^7$ | $3.872 \cdot 10^6$ | $3.871 \cdot 10^6$ | $5.770 \cdot 103$ | $5.768 \cdot 10^3$ |
| $Na^+$ | $4.680 \cdot 10^6$ | $4.674 \cdot 10^6$ | $1.170 \cdot 10^6$ | $1.169 \cdot 10^6$ | $1.743 \cdot 103$ | $1.741 \cdot 10^3$ |
| $K^+$ | $2.754 \cdot 10^6$ | $2.748 \cdot 10^6$ | $6.886 \cdot 10^5$ | $6.871 \cdot 10^5$ | $1.026 \cdot 103$ | $1.024 \cdot 10^3$ |
| $Rb^+$ | $1.263 \cdot 10^6$ | $1.257 \cdot 10^6$ | $3.158 \cdot 10^5$ | $3.143 \cdot 10^5$ | $470.540$ | $468.339$ |
| $Cs^+$ | $8.145 \cdot 10^5$ | $8.086 \cdot 10^5$ | $2.036 \cdot 10^5$ | $2.021 \cdot 10^5$ | $303.382$ | $301.181$ |
| $Be^{2+}$ | $1.193 \cdot 10^7$ | $1.193 \cdot 10^7$ | $2.983 \cdot 10^6$ | $2.982 \cdot 10^6$ | $4.445 \cdot 103$ | $4.443 \cdot 10^3$ |
| $Mg^{2+}$ | $4.426 \cdot 10^6$ | $4.420 \cdot 10^6$ | $1.107 \cdot 10^6$ | $1.105 \cdot 10^6$ | $1.649 \cdot 103$ | $1.647 \cdot 10^3$ |
| $Ca^{2+}$ | $2.687 \cdot 10^6$ | $2.681 \cdot 10^6$ | $6.718 \cdot 10^5$ | $6.703 \cdot 10^5$ | $1.001 \cdot 103$ | $998.700$ |
| $Sr^{2+}$ | $1.232 \cdot 10^6$ | $1.226 \cdot 10^6$ | $3.081 \cdot 10^5$ | $3.066 \cdot 10^5$ | $459.049$ | $456.847$ |
| $Ba^{2+}$ | $7.884 \cdot 10^5$ | $7.825 \cdot 10^5$ | $1.971 \cdot 10^5$ | $1.956 \cdot 10^5$ | $293.668$ | $291.467$ |
| $Al^{3+}$ | $3.989 \cdot 10^6$ | $3.983 \cdot 10^6$ | $9.972 \cdot 10^5$ | $9.957 \cdot 10^5$ | $1.486 \cdot 10^3$ | $1.484 \cdot 10^3$ |
| $Y^{3+}$ | $1.215 \cdot 10^6$ | $1.209 \cdot 10^6$ | $3.036 \cdot 10^5$ | $3.022 \cdot 10^5$ | $452.420$ | $450.219$ |
| $La^{3+}$ | $7.795 \cdot 10^5$ | $7.736 \cdot 10^5$ | $1.949 \cdot 10^5$ | $1.934 \cdot 10^5$ | $290.374$ | $288.173$ |

32 MHz and about 21 MHz for aquatic complexes formed by the $Y^{3+}$ and $La^{3+}$ cations, respectively. Table 2.2 gives hypothetical (expected) $v_p$ values obtained under various assumptions regarding the effective value of $r_0$ and effective mass of associated solvate groups.

Estimates show that the number of $OH_2$ solvate groups structured around the ion within the solvation shell does not have a decisive influence on the frequency value for the case when the mass of the hydration shell significantly exceeds the mass of the cation (Table 2.3). As can be seen from Fig. 2.6, the mass of the hydration shell exerts a significant effect on $v_p$ only up to a hydration shell mass

72

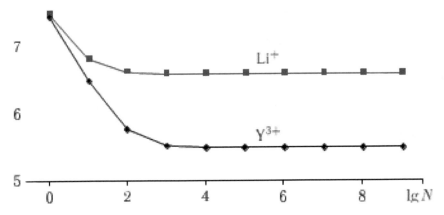

**Fig. 2.6.** The dependence of the frequency $v_p$ for yttrium and lithium cations on the number of water molecules in the aquacomplex on a logarithmic scale ($r_0$ is assumed to be 100 diameters of a water molecule).

equal to $10^4$ masses of water molecules for an aquacomplex based on a $Y^{3+}$ cation and $10^3$ masses of water molecules for an aquacomplex based on a $Li^+$ cation.

The size of the solvate shell, or rather, the supramolecular formation consisting of solvate groups, has a decisive influence on the $v_p$ value (Table 2.4).

**Table 2.3.** Dependence of the frequency $v_p$ on the number of associated water molecules (solvate groups) for $Y^{3+}$ and $Li^+$ cations ($r0 = 100d_{H_2O}$)

| Number of hydrate groupos $N$, number of water molecules | lg $N$ | $v_p$, Hz | |
|---|---|---|---|
| | | $Y^{3+}$ | $Li^+$ |
| $10^0$ | 0 | $2.72 \cdot 10_7$ | $3.07 \cdot 10^7$ |
| $10^1$ | 1 | $2.99 \cdot 10^6$ | $6.56 \cdot 10^6$ |
| $10^2$ | 2 | $5.71 \cdot 10^5$ | $4.14 \cdot 10^6$ |
| $10^3$ | 3 | $3.29 \cdot 10^5$ | $3.90 \cdot 10^6$ |
| $10^4$ | 4 | $3.05 \cdot 10^5$ | $3.87 \cdot 10^6$ |
| $10^5$ | 5 | $3.02 \cdot 10^5$ | $3.87 \cdot 10^6$ |
| $10^6$ | 6 | $3.02 \cdot 10^5$ | $3.87 \cdot 10^6$ |
| $10^7$ | 7 | $3.02 \cdot 10^5$ | $3.87 \cdot 10^6$ |
| $10^8$ | 8 | $3.02 \cdot 10^5$ | $3.87 \cdot 10^6$ |
| $10^9$ | 9 | $3.02 \cdot 10^5$ | $3.87 \cdot 10^6$ |

**Table 2.4.** The dependence of the characteristic frequencies on the size of the aquacomplex (the mass of the solvation shell is $10^4$ masses of a water molecule)

| $r_0$, m | Characteristic frequencies, kHz | |
|---|---|---|
| | Aquacomplex $Li^+$ | Aquacomplex $Y^{3+}$ |
| $1.93 \cdot 10^{-9}$ | $3.87 \cdot 10^5$ | $3.02 \cdot 10^4$ |
| $1.93 \cdot 10^{-8}$ | $3.87 \cdot 10^3$ | 302 |
| $3.86 \cdot 10^{-8}$ | 968 | 75.6 |
| $5.79 \cdot 10^{-8}$ | 430 | 33.6 |
| $7.72 \cdot 10^{-8}$ | 242 | 18.9 |
| $9.65 \cdot 10^{-8}$ | 155 | 12.1 |
| $1.16 \cdot 10^{-7}$ | 108 | 8.40 |
| $1.35 \cdot 10^{-7}$ | 79 | 6.17 |
| $1.54 \cdot 10^{-7}$ | 60.5 | 4.72 |
| $1.74 \cdot 10^{-7}$ | 47.8 | 3.73 |
| $1.93 \cdot 10^{-7}$ | 38.7 | 3.02 |

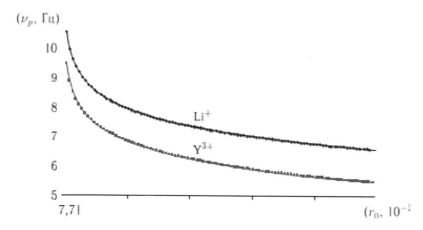

**Fig. 2.7.** Frequency dependence for yttrium and lithium cations on the cation– solvation shell distance; the mass of the solvate shell is $10^4$ masses of water molecules.

The presented dependences (Fig. 2.7) (as an example) demonstrate the behaviour of the characteristic frequencies for the light cation $Li^7$ and the heavier cation $Y^{89}$. As can be seen from the figure, the difference in the masses of $1/12.7$ provides a difference in the values of the characteristic frequencies by about an order of magnitude for different sizes of the aquacomplex. Characteristic frequencies are 3.02 and 38.7 kHz for aquatic complexes of cations $Y^{3+}$ and $Li^+$, respectively, with a radius of supramolecular formations of 0.0193 μm (100 diameters of a water molecule).

Thus, the calculation results show the following. The characteristic frequency of rotation of the solvate groups relative to the cation for a particular cation (nuclide) depends both on the mass of the solvate shell (the number of solvate groups) and on the size of the supramolecular formation as a whole. The dependence of $v_p$ on mass has an asymptote, which is determined by the mass of the cation and the size of the aquacomplex. The dependence of $v_p$ on the size of the supramolecular formation has no special features, and the characteristic frequency decreases with increasing cation mass.

This fact indicates that the implementation of the separation process of the elements in a solution of a mixture of their salts is less energy intensive for relatively heavy elements, when the 'working' frequency of the process is lower.

## 2.3. Mechanistic models of the ion–solvate shell system

The efficiency of the excitation of the phenomenon of selective drift of solvated cations of different metals in solutions of a mixture of salts under the action of an external periodic electric field increases significantly in rather narrow ranges of field frequencies and is probably resonant in nature. To determine the intervals of resonant frequencies, it is necessary to choose a convenient model of the ion–solvate shell system that makes it possible to carry out calculated estimates and serves as a starting point in the construction of more complex models that take into account subtle effects. In this paper, several possible models are presented and analyzed, one of which gives the result consistent with experimental results.

### 2.3.1. Solvated ion (cluster) as a drop of liquid

We will consider a cluster formed by a central ion and solvent molecules associated around it as a drop of liquid. For droplet oscillations relative to the static equilibrium position under the assumption that the perturbations of the velocity field are small inside the drop and the free surface, the natural frequencies can be determined by the formula [21]

$$v_l^2 = \frac{\sigma_s}{\rho R_0^3} l(l-2)(l+2), \quad l = 2,3,...,$$

where $\rho$ is the fluid density, $\sigma_s$ is the surface tension coefficient, $R_0$ is the radius of the drop (cluster).

To determine the surface tension coefficient, it is necessary to evaluate the work that needs to be done to create a surface bounding the cluster. The exit of the solvent molecule to this surface leads to an increase in the cluster energy by the binding energy of the solvent molecule inside the cluster. The coefficient of surface tension will be equal to the ratio of this energy to the area of the surface of the sphere per one molecule of the solvent:

$$\sigma_s = \frac{\varepsilon_b}{\varsigma}.$$

Theoretical and experimental studies of the processes of formation of nanometer structures in polar dielectric liquids under the influence of an electric field showed that due to the dipole–dipole interaction of water molecules, molecular bridges can be formed in the interelectrode gap. In this case, there is a critical electric field $E_{cr}$ for the formation of molecular bridges. For $E > E_{cr}$, the polarized molecules will be connected by a dipole–dipole interaction and are oriented in the direction of the field of the central ion. For $E < E_{cr}$, the thermal motion of the molecules must destroy the bridges. The cluster radius can be estimated from the condition $|E| = |E_{cr}|$ (see section 2.5). This condition means that for $r = r_{cl}$ the field of the central ion is 'balanced' by the oppositely directed field of the 'bridges' of the lined dipoles. In other words, a 'fur coat' from the lined dipoles shields the central ion.

To carry out the estimates, we will now consider the radius of the cluster $R_0$ equal to 0.2 μm ($E_r(r) = E_{cr}$). Under normal conditions, $9 \cdot 10^5$ water molecules are placed within a sphere of radius $R_0$. The surface area of the sphere is $50.24 \cdot 10^{-14}$ m$^2$, with $5.55 \cdot 10^{-19}$ m$^2$ per molecule. To determine the binding energy of the solvent molecule inside the cluster, the following can be suggested:

1) It is equal to the binding energy of the dipole in the electric field of the central ion,

$$\varepsilon_b \sim p \cdot E,$$

where $p$ is the intrinsic dipole moment of the solvent molecule, $E$ is the electric field strength within the radius of the cluster. For the intrinsic dipole moment of a water molecule $p = 6.2 \cdot 10^{-30}$ C $\cdot$ m, the binding energy is about $10^{-24}$ J. The surface tension coefficient in this case $\sigma_s \approx 2 \cdot 10^{-6}$ J/m$^2$. Then the natural vibration frequency of the cluster as a drop can reach several hundred kHz.

2) It is equal to the hydrogen binding energy in the solvent per one solvent molecule. Hydrogen binding energy in water is 21 kJ/mol, therefore, for one molecule of the solvent $\varepsilon_b \approx 3.5 \cdot 10^{-20}$ J. The surface coefficient tension in this case $\sigma_s \approx 0.063$ J/m$^2$. Then the natural frequency of the oscillations of the cluster as a drop will amount to tens of MHz.

### 2.3.2. Solvated ion (cluster) as a spherical pendulum

The model of a cluster formed by a central ion and solvent molecules associated around it as a hollow spherical shell is presented in Sec. 1.1.3. For the $Ca^{2+}$ ion, the radius $r_i = 0.114 \cdot 10^{-9}$ m with a coordination number of $n = 6$ and $r_i = 0.148 \cdot 10^{-9}$ m with the coordination number $n = 12$. In this case, the first solvation radius will be $R_1 = 0.229 \cdot 10^{-8}$ m for $n = 6$ and $R_1 = 0.324 \cdot 10^{-8}$ m for $n = 12$. If we assume that the outer radius $R_2$ is determined from the conditions $E_r(R_2) = E_{cr}$ and is 0.2 μm, then the mass of the torsion spherical pendulum will be equal to $M \approx 33.5 \cdot 10^{-18}$ kg. If the solvent is water, then the water molecules will be oriented in such a way that the oxygen atoms will be closer to the central ion of $Ca^{2+}$. Due to the significant difference in the electronegativity of hydrogen and oxygen atomic oxygen in a water molecule, a significant negative partial charge $q = -0.66e$ arises. As a result of the deviation of one water molecule from the equilibrium position in the solvate within the first radius of solvation for the solvated $Ca^{2+}$ ion, the stiffness value will be $k = 55 \cdot 10^{-6}$ N/m. Then the natural vibration frequency of the cluster as a torsion spherical pendulum will be about 7 THz.

### 2.3.3. The solvated ion (cluster) as a spherical rotator

We will consider a cluster formed by a central ion and solvent molecules associated around it as a hollow spherical shell with an internal radius $R_1$, an external radius $R_2$ ($R_2 \gg R_1$) and mass $M$. Inside the cavity is located a central ion.

The volume of the cavity is very small compared to the volume of the entire cluster, having a radius of $R_2$. In this case, the hollow spherical shell may rotate relative to the centre of mass of the cluster, which practically coincides with the geometric centre of the spherical shell, since the mass of the shell $M$ significantly exceeds the mass of the central ion $m_i$. The natural frequencies for such a system can be determined according to the formula [22]

$$v_l = \frac{h}{8\pi^2 l} l(l+1), \quad l=1,2,3,...,$$

where $h$ is the Planck constant, $J = MR_2^2$ is the moment of inertia of the rotator.

Depending on the value of $R_2$, which in turn determines the value of $M$, natural frequencies vary from units of Hz for the cluster stera with radius $R_2 \sim 0.1$ μm to 10 MHz for the cluster where the solvate shell of which is formed by 12 water molecules.

Solvation lead cation separation experiments and experiments with separation of cerium in nitric acid aqueous solutions under the action of an external electric field (see Section 1.3) at concentrations of metal salts of 0.1 g/$l$ under the conditions of electrical insulation from a solution of electrodes forming a field, show that the influence of the field with a frequency of 100 Hz at a tension of about 80 V/cm ensures stable separation of the cationic aquacomplexes. The maximum separation ratio is 1.054. Thus, to describe the process of excitation of the motives of rotational–translational motion of solvated ion under the influence of an external electric field, a model with a solvated ion (cluster) as a spherical rotator is suitable. The natural vibrational frequencies of an ion in a salt solution in a polar dielectric liquid are in the range from tens of Hz to units of kHz and correspond to two possible solvated models:

1) The mass of the solvation shell does not exceed a value of $10^{-22}$ kg, the outer radius of the shell does not exceed the value of 0.05 μm, which is an order of magnitude greater than the first radius of solvation. If the solvent is water, within the solvation shell there are up to $10^4$ water molecules. The natural frequencies of the rotator will be tens of Hz.

2) Only relatively mobile solvent molecules, not strongly coupled by the electric field of the central ion and directly adjacent, take part in the rotational motion to a surface with radius $R_2$. The value of $R_2$ is determined from the condition $E_r(R_2) = E_{cr}$. The total number of such mobile molecules is significantly smaller than the number of the solvent molecules associated around the central ion. The mass of such a spherical rotator formed by a layer of solvent molecules adjacent to the outer boundary of the solvation shell does not exceed $10^{-23}$ kg (350–400 water molecules). The natural frequencies of the rotator in this case will be units of kHz.

## 2.4. Solution structure and dimensions of solvated ions

When dissolved, complexation occurs, as a result of which supramolecular formations are formed – hydrated aquacomplexes in aqueous solutions. The formation of aquacomplexes is due to the interaction of cations (anions) with water molecules. Water complexes, in turn, are also hydrated. That is around each of them there are coordinated water molecules (solvate groups). The influence of the central ion on the surrounding water molecules is limited and it can be said that only those water molecules that are within a sphere of a certain radius with a centre coinciding with the centre of the ion, are affected by the ion. The aquacomplex size is a value equal to the radius of such a sphere. It turns out that the salt solution is structured, and the unit structure of the solution is the solvated ion (in the case of an aqueous solution, an aquacomplex) [23, 24].

The notion that liquids are not at all structureless substances has formed relatively recently (thirty years of the last century), when it became obvious that a continuous model of a liquid state that does not distinguish between a gas and a liquid and is not able to describe many (in particular, critical) phenomena. X-ray studies of liquids that began at the same time, showed that in liquids there is some order that the immediate environment of each molecules resembles (albeit more loose and mobile) packaging in crystals. Thus, the idea of the near structural order in liquids is proposed [25].

According to Fisher [26] and Eisenberg and Kozman [27] in liquids three types of structures can be distinguished according to their characteristic times.

1) The immediate or **I**-structure of the immediate environment is essential for fast processes with a characteristic time of $\sim 10^{-15}$ s. Usually such a structure can be observed in computer experiments. For example, the time step of the numerical integration of the equations of motion in the molecular dynamics method $\Delta \tau \leq 10^{-15}$ s.

2) Vibrationally averaged or **V**-structure with a characteristic time varying $\sim 10^{-12}-10^{-13}$ s. It is the concept of **V**-structure that underlie the construction of various phenomenological models of the structure of water.

3) The diffusion-averaged or **D**-structure described by the molecular distribution functions that make up a new theoretical tool of the modern theory of liquids [28–30].

Very useful and illustrative is the idea of the **F**-structure, or hidden inherent structure (inherent hidden structure), introduced by

Weber and Stilinger [31]. This structure corresponds to the minimum potential energy to the nearest instantaneous **I**-structure.

Water refers to 'difficult' liquids in which the interaction between molecules is no longer described by spherically symmetric functions, but also depends on the orientation of the molecules relative to each other, which leads to a complex orientation dependence of the distribution functions. New problems arise in the description of hydrogen bonds between water molecules. It is thanks to these bonds in water that cause a unique tetrahedral short-range order with which all the unusual properties of water that distinguish it from other liquids arise, and the unusual properties of a huge number of different aqueous crystalline structures.

Conventionally, all structural models of water can be divided into two classed. This is a class of discrete Bernal and Fowler models and, for example, the well-known cluster model of Nemethy and Sheraga [32], which is a two-structure model of five states. At the heart of the model it is assumed that water molecules are either in compact structures – clusters with the number of hydrogen bonds 1–4 in the molecule –, or exist as monomers not bound by a hydrogen bond. These two structures are equilibrium mixed in accordance with the minimum free energy. The model is based on the assumption of the cooperative formation and destruction of hydrogen bonds, according to which the formation of a single hydrogen bond (due to local energy fluctuations) lowers the potential barrier for the formation of the next hydrogen bond, which leads to cascade cluster formation. Similarly, the reverse cooperative process occurs. The breaking of one hydrogen bond entails destruction of the entire cluster. The emergence and destruction of clusters is ongoing. The maximum density is due to the presence of two factors. On the one hand, since when melting in water a certain amount of unbroken hydrogen bonds is stored, then a further increase in temperature leads to their gradual destruction and an increase in the coordination number, i.e., local density. On the other hand, with increasing temperature, the usual expansion mechanism inherent in all liquids also occurs. Up to a temperature of 4°C, the first mechanism dominates, and then the second. Such the interpretation of the maximum density is inherent to one degree or another in all existing water models.

Note another, somewhat exotic clathrate Polling model [33], whose element is the dodecahedron formed by hydrogen-bonded molecules in the cavity of which there is unbound water. The dodecahedrons themselves are also interconnected, forming some spatial network.

This model is very close to Samoilov's partially filling model [34]. It is based on the idea that water molecules form a spoiled blurry ice I-structure with a partial filling of the structure cavities with monomers, and the lattice is constantly reconstructed during the movement of the molecules. The existence of a maximum density in water is explained by the fact that, as the temperature increases from 0∘C, on the one hand, vibrations of molecules near the equilibrium positions in the structure becomes stronger and, accordingly, the effective radius of the molecule grows, and on the other hand, the translational movement intensifies and more and more molecules fall into the void. The first circumstance leads to an increase in volume, the second to compaction.

The so-called *continuous model of Pople's curved bonds* [35], in which the hydrogen bond is described only by the electrostatic interaction between protons of one molecules and a lone pair of electrons second. In fact, the entire system is connected by a flexible and extensible network of electrostatic interactions. Similar mesh models are still successfully used. in various physicochemical applications [35, 36] for the reason that many properties of liquids, such as diffusion, compressibility, etc., can be adequately interpreted as properties of a model network.

The modern theory of liquids (in principle) is capable of solving structural problems in the region of the existence of stable water, i.e., at temperatures above 0°C. However, in the field of metastable and, in particular, amorphous states of water, the problems facing the theory so far seem difficult to overcome.

Structural theories of ionic solvation, constructed in the 80–90s of the last century, as a rule, were based on the consideration of the interactions of ions with their environment in the solution as individual formations, and not as stoichiometric mixtures of the opposite charged particles. The quality of these theoretical approximations was often evaluated by comparing the calculated data

**Table 2.5.** The values of the surface potentials of solvents

| Solvent | $\chi$, V | Solvent | $\chi$, V |
|---|---|---|---|
| Water | 0.100 | *n*-butyl alcohol | −0.277 |
| Methyl alcohol | −0.184 | Acetonitrile | −0.108 |
| Ethyl alcohol | −0.260 | Dimethylsulfoxide | −0.238 |
| *n*-propyl alcohol | −0.267 | Dimethylformamide | 0.434 |
| Isopropyl alcohol | −0.275 | Acetone | −0.337 |

with the 'experimental' ones obtained using various kinds of extra-thermodynamic assumptions based on the use of model compounds.

One of the ways to resolve this fundamental dilemma, according to the authors of [37], is to use the concept of real thermodynamic properties of individual ions, which is based on taking into account interfacial potentials in calculating the energy values of Gibbs ion solvation from experimental data on volt potentials [38, 39].

The value of the chemical energy of solvation is determined only by the difference in the state of ions inside two phases – vacuum and solution. However, in order to calculate the changes in the Gibbs energy upon solvation of an individual ion, it becomes necessary to take into account the work of the transition of the ion through the phase boundary [38–40]. The gas/solvent interfacial surface phases are preferably formed by oriented dipoles of polar solvent molecules, which leads to a predominant charge of the distribution of one or another sign. As a result, a potential difference arises on the surface of the solvent or simply the surface potential $\chi$. Therefore, the work on the transfer of cations and anions through the interfacial layer will differ. So, if the surface is positively charged, the entry of cations will be accompanied by energy costs.

The exact determination of $\chi$ is also one of the important points in solving the problem of calculating the chemical thermodynamic characteristics of the solvation of individual ions. The correctness of subsequent calculations and the reliability of conclusions to a large extent depend on how correctly the choice of the numerical value of $\chi(H_2O)$ as a reference value is made.

The analysis of the available data carried out in [41–43] showed that the value $\chi(H_2O) = 0.10V$ is the most reliable. Using this value, the $\chi(S)$ values of a number of organic solvents were determined. They are given in Table 2.5.

As follows from the data presented, the values of the surface potentials of organic solvents have negative values. This allows us to conclude that the molecules of these solvents are oriented in such a way that the polar functional groups in the surface layer are directed into the depth of the liquid phase, and nonpolar radicals are directed to the side of the gas phase.

According to the ideas developed in [44–46], one of the stages of solvation of an ion is the neutralization of its charge and the redistribution of the ion over the surrounding solvent molecules. As a result, the ion in solution appears as a kind of hypothetical neutral particle, isoelectronic to the corresponding atom of a noble gas.

Generally speaking, the fact that the main quantity determining the size of the solvated ion, which is a supramolecular formation – a 'cluster' – is the electric potential and its distribution around the ion (cation or anion) located in a continuous environment formed by polar molecules of water follows from general physical reasoning.

The basis of modern concepts of the interaction of solved compounds with a solvent are the classic works of Born, Debye, Lennard–Jones, London, Kirkwood and Onsager. These works, their further development and role in chemistry are analyzed in detail in the monograph [47]. However, the results presented there for accounting for solvation are not successful, therefore, simpler models have been developed that allow us to calculate the solvation energy with a fairly high degree of accuracy. Before proceeding to the description of these models, we consider the possible types of solvation.

Solvation is usually divided into two types – specific and non-specific. The former is characterized by the existence in solutions of structurally defined formations between the solvate and the solvent, the lifetime of which significantly exceeds the period of free oscillation of this isolated system. Examples of such formations are hydrogen bonds, charge transfer complexes, etc. All of them are characterized by structural certainty and a relatively large value of the interaction energy, which can sometimes exceed 10% of the chemical bond energy, as well as a significant redistribution of the charge between the dissolved compound and the solvent.

A special case is the solvation of small single di- and some tri- and tetraatomic ions. These compounds are characterized by all of the above features, but the bond is unusually strong. So, the energy of hydration $OH^-$ is 445 kJ/mol, $H_3O^+$ 382 kJ/mol, $NH_3$ 344 kJ/mol. Such structures, apparently, would be more correct to consider as coordinated.

In the case of nonspecific solvation, the interaction energy between the dissolved compound and solvent molecules is small and cannot create solvate shells with a rigidly fixed structure. It is believed that the main contribution to the energy of solvation of this type is made by the van der Waals interaction and electrostatic interaction of the dipole moments of the dissolved compound and the solvent molecules, although it is generally accepted that dissolution involves dissociation of the molecules of the dissolved substance on ions. There is some contradiction, since there is no point in speaking about the dipole moments of the dissolved compound.

It is correct to assume that the main contribution to the solvation energy is made by the electrostatic interaction of the ions of the dissolved compound and the dipole moments of the solvent molecules. In this case, the formation of solvate shells with a rigidly fixed structure is quite natural.

The methods of quantum chemistry, which are currently used to describe solvation effects, can be divided into two large groups: discrete and continuum. Discrete approaches usually deal with the description of any allocated volume of the 'dissolved compound–solvent' system with the inclusion of up to ten molecules of the latter. These molecules are distributed around a circle of the dissolved compound in a certain system (approximation of the 'supermolecule', various versions of the model of point dipoles or charges). A similar description is close to the chemical determination of specific solvation. They differ only in that the nonspecific interaction of the dissolved compounds and solvent are taken into account to a certain extent in the calculations.

Some papers considered discrete systems containing 200–300 solvent molecules (Monte Carlo method with atomic–atomic potentials). Such large clusters are suitable for describing both specific and nonspecific interactions of a dissolved compound with a solvent. However, the calculations for them are associated with a very large expenditure of the processor time. Therefore, to take into account nonspecific solvation, continuum models (solvaton model, self-harmonized reaction field) are used extensively. They do not take into account the microscopic structure of the solvent, so there is no need to determine the structure of the solvate shell. Thanks to this, calculations becomes much easier.

The total energy of a molecule in a solvent can be represented as the sum of two separate contributions:

$$E_{ful} = E_m + E_s$$

where $E_m$ is the energy of an isolated molecule; $E_s$ is the solvation energy.

Under the assumption that there is no specific interaction between the dissolved compound and the solvent, i.e., a substantial redistribution of charge between them (this case will be consideredseparately), $E_s$ is the sum of three contributions:

$$E_s = E_{el} + E_{disp} + E_{cav}$$

where $E_{el}$ is the energy of the electrostatic interaction between the intrinsic and induced charges of the dissolved compound and the solvent molecules; $E_{disp}$ is the dispersion component of the interaction energy, taking into account the van der Waals interaction; $E_{cav}$ is the so-called *cavitational energy*, i.e., the energy of reorganization of the solvent necessary for the formation of a cavity in which a dissolved compound is placed. Calculations taking into account all three contributions to solvation energy showed that for the most interesting polar and charged systems in polar solvents, when solvation can significantly change the results of gas-phase calculations, $E_{el}$ exceeds $E_{disp}$ and $E_{cav}$; in addition, it turned out that it was $E_{el}$ that changes significantly during the reaction, and there were small changes in $E_{disp}$ and $E_{cav}$ which to a large extent cancel each other out. It led to the development of a number of models in which solvation is taken into account in the electrostatic approximation (various versions of the models of point dipoles or charges, most continuum models).

Currently, many different methods for accounting for solvation are proposed. Below we will consider only those that are received most widely.

Jano [48] and Klopman [49] proposed the *Solvaton model*. It is based on the assumption that, with each atom of the dissolved compound there is an electric charge induced in the solvent – solvaton. The solvaton charge in the absolute value is equal to the charge on the atom that induced it, but has the opposite sign. The solvatons do not interact with each other. Energy interactions between solvaton $C$ and atom $A$ are calculated by the following formula:

$$E_{CA} = q_C q_A e^2 (\varepsilon - 1)/(2r\varepsilon),$$

where $r$ is the radius of atom $A$, if solvaton $C$ is induced by a charge of the given atom, or the distance between atoms $A$ and $B$, if the solvaton is induced by the charge of atom $B$; $q_C$ and $q_A$ are solvaton $C$ and atom $A$ charges in the units of electron charge; $e$ is the electron charge; $\varepsilon$ is the dielectric constant of the medium.

Most often, for the evaluation of $E_{CA}$, a similar expression is used with Coulomb's integrals taken from the quantum chemical calculation, i.e., $e^2/r$ is set equal to $r_{AA}$ if solvaton $C$ is induced by near atom $A$, or $r_{AB}$, if solvaton $C$ is induced by a charge of the atom $B$ ($r_{AA}$ and $r_{AB}$ are one- and two-centre Coulomb integrals, respectively). For the solvation energy in this approximation, we can write the following expression, where the sum is taken over all atoms of the dissolved molecule:

$$E_C = \sum_{A,B} q_A q_B \left[ (\varepsilon - 1) / 2\varepsilon \right] \gamma_{AB}.$$

The solvaton model is simple and economical. However the model possesses a number of serious flaws. The most unpleasant of them is that the solvatons are speculatively localized in the immediate proximity to the atoms of the dissolved molecule that induced them, that is, it is actually assumed that the solvatons penetrate inside the molecules. In the case of the ions, the solvaton model takes into account the change in the solvation energy only due to a change in the degree of delocalization of the charge, although in reality the determining factor in this case is their volume. If the change in the degree of delocalization of the charge correlates with a change in the volume of the ion, then the solvaton model will give the correct results. From this limitations of its scope become clear. In particular, the solvaton model cannot take into account the effects of charge screening by nonpolar groups or reagent molecules. For this reason, it cannot be used in combination with the approach of the 'supermolecule' if it is necessary to take into account the effect of specific solvation.

It is necessary to note one more important circumstance, consisting in that the charges on the atoms calculated by different methods strongly vary. Because of this, the solvaton model, being applied in the framework of different methods, gives different results. If the model is good when works with one method, then in calculations by other methods the results can be much worse.

In the framework of the macroscopic theory (a ball with a fixed charge distribution is immersed in a medium with a dielectric constant $\varepsilon$, charges are localized inside the ball), the interaction energy with the solvent is determined by the classical electrostatic potential

$$U = \sum_{i=0}^{\infty} U_i,$$

where $U_0 = (q^2/2\alpha)(1 - 1/\varepsilon)$, $U_1 = (\mu^2/2\alpha^3)[2(\varepsilon - 1)/(2\varepsilon + 1)]$, etc.

Here $q$, $\mu$ are the charge and dipole moment of the ball, respectively. The first member of this series is called Born, the second – Onsager. The following members of the series depend on quadrupole and higher multipole moments of the ball. For electrically neutral balls, $U_0 = 0$, and the first nonzero member of the series will be $U_1$. Tapia et al. [50–54] proposed adding an analogue of the term $U_1$ to the Hamiltonian isolated molecule and take into account the interaction in this way with the solvent. As a result, they obtained the following expression for calculating the Hamiltonian of a molecule in the solvent

$$\hat{H} = \hat{H}_0 - \mu g \langle \psi | \mu \psi \rangle,$$

where $g$ is the tensor, which is determined by the susceptibility of the electrostatic field of the molecule to the medium. In [50–54] it was given in the parametric form.

This method of taking into account the interaction with the solvent is called the self-consistent reaction field model. Its main disadvantages are obvious. They consist in the need to take into account the dependence of tensor $g$ on the value of the dielectric constant of the solvent and the size of the molecule. This cannot be done without very crude additional suggestions. It is usually assumed that the diagonal elements of $g$ are $2(\varepsilon - 1)/[(2\varepsilon + 1)a^3]$ and off-diagonal are equal to zero. However, the transfer of a macroscopic formula to a microscopic level, generally speaking, is wrong, since the concept of dielectric permeability in this case loses its physical meaning, and the choice of the value of parameter $b$ (the radius of the molecule) is quite arbitrary. therefore the self-consistent reaction field model has a limited region of application. In practice, it can be used when considering reactions of isomerization and proton transfer. In other cases, its use is undesirable.

From the point of view of building a model, the easiest way taking into account solvation is the inclusion of a large number of molecules of the medium in a system for which a quantum-chemical calculation is performed. All electrons of such a 'supermolecule' (solvated molecule and medium molecules) are included in the electronic Hamiltonian. This way is a direct generalization of quantum chemical methods developed for individual molecules in the case of large systems consisting of several or even a large number of individual molecules. If it is possible to take into account in this way the interaction of dissolved molecules with a large number of molecules of the medium, calculate energetically the most favourable conformation of the dissolved molecule and the configuration of the solvent molecules and get for this configuration electronic wave function, it can be explained or even predict almost all properties of a molecule of interest in the solvent. In this case, it is necessary ti with extreme caution to approach the choice of the quantum chemical method, to which in this case it is essential to present increased requirements, namely: it should be suitable for the study of intermolecular interactions.

Many quantum chemical methods that are successfully used to study the reactivity of organic compounds give wrong results when calculating the parameters characterizing the intermolecular interactions. Most often, unsatisfactory results are obtained when calculating systems with hydrogen bonds. Therefore, when using the 'supermolecule' approximation to take into account solvation, one has to choose a sufficiently perfect quantum chemical method. In this case, difficulties arise with a very large size of a supermolecule, which consists of a dissolved compound and molecules of the medium. Also in the supermolecule usually a large number of geometric parameters that determine the structure of the solvate shell are not known. Therefore, in reality, such calculations can be carried out successfully only for systems with a small number molecules of the environment.

Such accounting for solvation, in which the system from a solvated molecule and some limited number of molecules of the solvent is calculated by the quantum chemical method as one molecule, received the name 'supermolecular approximation' [55, 56]. Calculations in this approximation are widespread. These include work where the parameters are calculated for complexes consisting of a dissolved molecule and one solvent molecule, as well as calculations for complexes with two, three, etc. molecules of the medium. In these studies it was possible to obtain for the first time data on the structure and interaction energy for solvate shells of $OH^-$ and $H_3O^+$ ions. Some time later, solvation shells of the simplest ions were studied by similar methods: $Li^+$, $Be^{2+}$, $Na^+$, $Mg^{2+}$, $Al^{3+}$, $K^+$, $Ca^{2+}$, $F^-$, $Cl^-$, $NH_4^+$ [57–64], $CH_5^+$, $CH_5^-$ [65], alkyl ammonium ions [66].

The methodology for the above work laid the foundation of the approach to the study of solvation in the 'supermolecule' approximation. The essence of these approaches is as follows. First, the interaction energy of an ion with one water molecule is calculated and the most energetically favourable conformation is determined, then another water molecule is added to the system and calculated again are the interaction energy and complex structure, etc. As a result such a calculation yields a set of quantities that are energies of hydration of ion $A$ with each subsequent water molecule ($E_{n,\,n-1}$):

$$A(n-1)H_2O + H_2O \rightarrow AnH_2O + E_{n,\,n-1}A.$$

For the simplest ions, these energies were compared with the experimental data obtained by high pressure mass spectrometry and cyclotron resonance methods. Good agreement with the experiment confirmed the wide possibilities of quantum-chemical calculations for studying solvation. In addition, data were received on the number of water molecules in the first sphere of the solvate shell of their configuration.

When moving to more complex molecules, calculations in the 'supermolecule' approximation are significantly more complicated, since there are already a large number of centres of solvation and it is not possible to fill completely the entire first solvation shell. Therefore, in most works the following sequence are adhered to:

1) the interaction of the solvated molecule is studied in detailwith one molecule of solvent and preliminary understanding of the structure of the solvate shell;

2) the effect of the second, third, etc. solvent molecules is studied on the structure of the solvate shell and the data obtained in the first stage are specified;

3) the maximum possible number of solvent molecules is introduced into the system under study, for which it is possible to perform a calculation, and for such a supermolecule, all interesting quantities are calculated.

A detailed study of the interaction of solvated compounds and one solvent molecule includes their approach (with different relative orientation) and rotation around their own local axes. At this stage, data are obtained on the main most profitable positions of the solvent molecules, the interaction energies at such positions, the distances between the dissolved compound and the solvent molecules, and the mobility of the latter. Usually three types of mobility are specified: in equilibrium, near the equilibrium position and far from the equilibrium position.

*Mobility in equilibrium position* – the possibility of a molecule solvent rotate around its own local axes. For values of this type of mobility it is necessary to calculate the dependence of the energy of the interaction of the dissolved compound and the solvent molecule on the angles of rotation of the latter around its local axes. Distance between they do not change. Mobility near the equilibrium position - the ability of the solvent molecule to move small distances from the equilibrium position. To study this type the mobilities calculate the form of the potential energy surface (PES) at small distances of the solvent molecule from the equilibrium position without breaking hydrogen bonds, which were in equilibrium.

The potential energy surface (PES) is the potential function (potential) of the interaction of atomic nuclei in an isolated molecule or chemical system consisting of interacting atoms and (or) molecules. A system containing $N$ atoms generally has $(z-3N-6)$ internal degrees of freedom $q_i$ ($i = 1, 2, ..., z$), which can be selected in various ways. The potential $U$ of the nuclei of atoms (i.e. PES) is a function of these degrees of freedom: $U = U(q_i)$. It enters the equation of motion (evolution) of the system and along with the operator of the kinetic energy of nuclei $T$ is the nuclear Hamiltonian $H_{nuc}$

$$H_{nuc}(q_i) = T + U(q_i). \qquad (2.12)$$

In the quantum-chemical calculation of the PES $U_k(q_i)$ for each electronic state, find the solution of the Schrödinger equation

$$[H_{el}(x_n|q_i) - U_k(q_i)] \, \Psi_k(x_n|q_i) = 0, \qquad (2.13)$$

in which the electron Hamiltonian $H_{el}$ and the electron wave function $\Psi_k$ depend on the coordinates of electrons $x_n$ (spatial and spin) as on variables, and the coordinates of the nuclei are parameters. In equation (2.13), the difference between $x_n$ and $q_i$ is indicated by the vertical line. Equation (2.13) is solved many times for different sets of parameters $q_i$, that is, for various fixed nuclear configurations. The resulting PES $U_k(q_i)$ is called an electronic term (usually if the quantum number $k$ of the electronic state is not indicated, the name 'PES' refers to the ground state $k = 0$, i.e., $U(q_i) = U_0(q_i)$).

Mobility away from the equilibrium position is considered as the ability of the solvent molecule to move long distances from the equilibrium position. To study this type of mobility, PES are calculated at large distances of the solvent molecule from the equilibrium position when the hydrogen bonds between it and the dissolved compounds are broken.

At the second stage, the number of solvent molecules in the supermolecule is increased. In this case, data on the positions of local minima obtained at the first stage are used and their position is only specify. In the third stage, they include as much as possible in the supermolecule the possible number of solvent molecules (based on the capabilities of the computer) and one or another parameter is calculated.

The approach of the 'supermolecule' allows us to solve many problems, associated with the effect of solvation on the reactivity of organic compounds. However, when using it, the most serious attention must be paid to the choice of calculation method. In this case, two requirements should be guided by: 1) the calculation method should be accurate enough and convey well both the basic properties of the dissolved molecules and the structure of solvate shells; 2) calculation in the approximation of a 'supermolecule' is associated with the calculation of the electronic wave function for a very large system and it should be practically implement based on the capabilities of existing computers, their speed and memory.

In the 'supermolecule' approximation, we considered a system of solvated compounds and a number of solvent molecules as one large molecule. This approach is a direct generalization of quantum chemistry methods developed for calculating the properties of individual (isolated) compounds on intermolecular interactions. Moreover, as the initial particles it is necessary to operate with electrons and atomic nuclei. When studying the system, consisting of one molecule, this approach is uniquely possible, since only at this level it is possible to analyze most chemical properties of a molecule. When considering intermolecular interactions, it becomes possible to operate not with electrons and atomic nuclei, but with individual molecules. For this, it is necessary to have potentials that describe intermolecular interactions. In the case of a system consisting of electrons and nuclei, it is necessary to solve the Schrödinger equation, since the electrons should be considered as quantum particles. When revising the intermolecular interactions the molecules can be considered as classic objects. Thanks to this, it becomes possible to use empirical potential functions to describe them. This substantially simplifies the task.

The various empirical potentials that were proposed by various authors to describe intermolecular interactions and do not satisfy the accuracy required when accounting for solvation. The lack of sufficiently reliable potentials made it impossible to use this approach to study solvation. Significant progress in this area was achieved through Clementi's et al. work. They put forward the idea of using non-empirical quantum-chemical calculations to determine potentials. intermolecular interactions [67–70]. Initially, potentials were generated numerically by non-empirical calculation of the interaction energy of a solvated molecule and a molecule of the solvent. But, since the further use of the numerical potential for constructing the solvate shell is difficult, an analytical function was selected for its approximation, which represented the sum of atomic-atomic potentials. All the atoms in the molecule were divided into classes depending on to which functional groups and in what positions in the group is a given atom. As a result, the number of classes many times exceeded the number of different atoms. For atoms of each class their atomic–atomic potentials were selected. The analytical form in which the search for atomic atomic potentials was performed was different and depended on the basis used in the calculation. When calculating the interaction potential between the molecules in small bases relatively simple analytic function was used:

$$U_{ij} = -\frac{A_{ij}}{r_{ij}^6} + \frac{B_{ij}}{r_{ij}^{12}} + \frac{C_{ij}q_iq_j}{r_{ij}}, \qquad (2.13a)$$

where $U_{ij}$ is the interaction energy between the volumes $i$ and $j$; $r_{ij}$ is the distance between these atoms; $q_i$ and $q_j$ are charges on atoms; $A_{ij}$, $B_{ij}$ and $C_{ij}$ are empirical parameters depending on which classes the atoms $i$ and $j$ belong.

To find the potentials of intermolecular interactions, non-empirical methods using large bases close to the Hartree–Fock method, were used to find more complex analytical functions. Calculations in large bases were carried out to determine analytical potentials describing the interaction between the water molecules. The calculations in the minimum basis were used to determine atomic-atomic potentials, describing the interact between the water molecule and the DNA bases, amino acids, etc.

The number of classes of atoms such as DNA and amino acids was several tens; the number of unknown parameters in atomic–atomic potentials reached several hundred. When determining the values of these parameters, one has to vary the relative position and mutual orientation of the molecules in sufficiently wide limits, almost to calculate each parameter 15–20 calculations were necessary. Thus, for calculating the intermolecular interaction potential of medium-sized molecules of the base type DNA and water molecules tens of thousands of the calculations of the total energy of the system by the non-empirical method are required. Therefore, the procedure for selecting parameters is associated with a very large expenditure of machine time. But one favourable circumstance should be noted: as parameter sets are accumulates and create them for each new connection, the amount of computation is reduced, since it is possible to attribute most of the atoms to already known classes for which all parameters of analytical potentials are known from calculations of other molecules [71].

After finding the potentials, the calculation of the structure of the solvate shell and the interaction energy between the solvent and the dissolved compound becomes a relatively simple task similar to conformational analysis tasks. The analytical form in which the potentials have been found to describe the interaction of the molecules of medium size and water molecules, also coincides with more widely used potentials used in the conformational calculations. However, the parameters in the Clementi potentials for intermolecular interactions are of a completely different nature. In conformational analysis, potentials of the type (2.13a) describe the van der Waals interactions between the atoms, and according to Clementi – electron-donor and electron acceptor interactions. The third term in the formula for atomic–atomic potentials corresponds to the Coulomb interaction. For electrically neutral molecules, the value of the coefficient $C_{ij}$ in the Clementi potentials is close to unity. However, for the ions it does not exceed 0.5; this is apparently due to the effects of screening and charge redistribution.

Using Clementi potentials allows us to consider hydration of very complex molecules by a large number of water molecules. So far, the number of water molecules does not exceed 10–15 no significant difficulties arise in calculating the structure of the solvation shell. However, with a further increase in the number of water molecules, a number of new problems appear. For a sufficiently accurate description of the solvation shells of even a small compound, it is desirable to increase the number of water molecules to 200–300. When calculating the structure of such a huge solvation shell the main difficulty is the existence of a large number of structures with close energies. The task is to find all such structures, determine the probability of implementation of each of them and carry out averaging over all found structures. In such calculations the temperature dependence must be taken into account..

In the works of Clementi it was shown that to find the structure of solvate shells, one can successfully use the Monte Carlo method, with the help of which the structure of solvate shells was used to calculate for a number of simple ions, taking into account their interaction with 200–250 molecules of water. In this case, another problem arose. Water molecule distribution in Monte Carlo calculations is probabilistic, therefore, the researchers faced the challenge of moving to such simple and intuitive hydration characteristics as the number of molecules water in the first sphere of the solvation shell and its radius. For getting this information it was proposed to calculate the dependence of the density of hydrogen or oxygen atoms on the distance to the centre of the ion. Such graphs get a series of pronounced highs. Their position for oxygen atoms is usually associated with the radii of the solvate shells, and the area under the curves is associated with the number of water molecules in the shell. Below are the radii of the first shells of the solvate spheres ($R$) and the number of water molecules in them ($N$) calculated by such a method [69, 70].

| Ion | $R$, nm | $N$ |
|---|---|---|
| $Li^+$ | 0.19–0.20 | 4 |
| $Na^+$ | 0.23–0.24 | 5–6 |
| $K^+$ | 0.28–0.29 | 5–7 |
| $F^-$ | 0.27–0.28 | 4–6 |
| $Cl^-$ | 0.34–0.35 | 6–7 |

The use of atomic–atomic potentials is very promising and can significantly expand our ideas about solvation and its effect on the reactivity of organic compounds. The bank of parameters is currently quite large, and it is hoped that in the future it will be further expanded. However, the assumptions made in these calculations should be emphasized.

1. The number of molecules in the solvation shell is not determined as a result of quantum-chemical calculations, and is set as the initial parameter for these calculations. This assumption can lead to the fact that the actual causes and effects in the process of solvation will be rearranged.

2. The approximate quantum-chemical method is used (very rough for medium-sized molecules) to calculate the parameters of the

atomic–atomic potentials. In the case of a small number of solvent molecules, the errors can be small, but as their number increases, they will accumulate.

3. Errors in the calculations may occur due to the approximation of the numerical potential by very simple analytic functions.

4. Atomic–atomic potentials, which are usually used to study solvation, are not additive functions, and the interactions of three bodies are rather difficult to take into account, they almost never do this, although these collective interactions significantly affect the calculation results (this is, apparently, the biggest drawback of the Clementi method).

In a number of works, Clementi's approach to accounting for solvation was used to study the effect of solvent on the potential surface energies of organic reactions [72, 73]. Carrying out such calculations requires a very long processor time. Their order is as follows:

1) calculate the total energy for any point on the surface of the potential energy of the gas-phase reaction;

2) at this point, calculate the parameters of atomic–atomic potentials that describe interaction of reagents with a solvent molecule;

3) using atomic–atomic potentials obtained at the previous stage of calculation (see paragraph 2), the solvation energy is calculated by the Monte Carlo method.

Such a chain of calculations has to be carried out for each point. of the potential energy surface, since during the reaction the electronic structure of the reactants changes significantly, which leads to a change in the parameters of the empirical potential function that describes interaction with a solvent molecule. Because of this, one cannot use a bank of ready-made parameters for atomic–atomic potentials; moreover, they have to be recounted at each new point surfaces of potential energy. This stage of calculation is connected with a very large amount of computation.

The main drawback of the 'supermolecule' approximation and Monte Carlo methods with atomic–atomic potentials lies in the extreme complexity of the calculation. Therefore, it was completely natural to develop methods that preserve the principle of these approaches, i.e., explicitly take into account the discrete set of solvent molecules around dissolved compounds, but modelled using point dipoles. In this approximation, the effect of an external electric field created by the solvent on the dissolved compound leads to the appearance of certain additional members that should be added the to the matrix elements of the Hamiltonian.

The point dipole model has been widely used by various authors to take into account solvation. However, almost all of these works had one significant drawback – they did not take into account the van der Waals repulsion between the dissolved compound and molecules of the solvent and the solvent molecules with each other. Because of this was it is impossible to calculate the geometry of the solvate shell, and the point dipoles simulating solvent molecules were arranged on new intuitive considerations.

A sequential electrostatic model that takes into account repulsion in a potential describing intermolecular interaction, was proposed by Worshel [74]. Molecules of the solvent in it were modelled by balls with a fixed dipole moment and the van der Waals radius. If the amount of the moleculesthere in the solvent is small, then using the Worschel method is straightforward. However, with an increase in the number of solvent molecules, difficulties arise associated with finding the optimal structure of the solvate shell. Using the Monte Carlo method for this purpose is associated with a very large amount of computation and not suitable for solving applied problems, and geometrical optimization methods of finding steepest descent allow one to find only one of many localized minima (not necessarily the deepest). Therefore, Worschel's method is currently rarely used.

Another version of the point dipole model was proposed in [75]. To avoid calculating the optimal solvate structure shell, the dipole moments of the solvent molecules were fragmented and almost continuously and uniformly 'smeared' in the solvent volume, i.e., each molecule of the solvent was replaced by a large number of point dipoles with small dipole moments. Such 'smearing' of the dipole moment of the solvent molecules earlier was used by Worshel in constructing the Langevin model dipoles [74].

Point dipoles with small dipole moments can be located at the nodes of any ordered lattice. From them follows highlighting the part of point dipoles that make a significant contribution to the energy of solvation. The position in space of this part point dipoles must satisfy the following two conditions. First, a point dipole cannot be closer than a certain critical distance to any of the atoms of a dissolved molecule. Secondly, the interaction energy between the point dipoles and a dissolved molecule must exceed a certain threshold value. The physical meaning of the first condition is obvious (point dipoles should not get inside the dissolved compound).

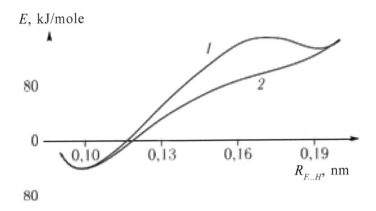

**Fig. 2.8.** PPE cross section for proton transfer reaction: 1 – distance between atoms F and N are fixed; 2 – distance between atoms $F$ and $N$ are optimized.

The essence of the second condition is to take into account the interaction of the dissolved compound only with those point dipoles that are in the region of a sufficiently strong electric field of the dissolved compounds. Their dipole moments will be oriented in the direction of this field, and therefore they will make the main contribution to the energy of solvation. Due to the interactions of the solvent molecules with each other the dipole moments of the remaining point dipoles will be oriented in arbitrary directions, therefore, the energy of their interaction with the electric field of the dissolved compound will be small. In [75] this contribution is proposed to be neglected.

The model of point dipoles is suitable only for a sufficiently rough qualitative account of solvation effects. Its main advantage is the low cost of processor time, therefore it is easy to change in applied calculations to obtain qualitatively correct results for reactions in polar solvents.

Let us consider the mechanism of dissociation of molecules in polar solvents. Experimental chemists often deal with such processes when studying many organic reactions, therefore, it is necessary to be able to model them in quantum chemical calculations. At first glance, this task seems very simple, but a closer acquaintance with it reveals that the mechanism of these processes is very complicated and generally accepted ideas about it are inaccurate

The processes are very complex and generally accepted ideas about it are inaccurate. In quantum-chemical studies, the formation

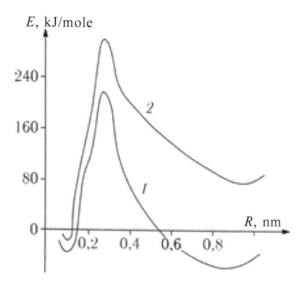

**Fig. 2.9.** The PES cross section for proton transfer reactions: 1 – reaction; 2 – $H_2O...H_2O \rightarrow OH^- + H_3O^+$. $R$ is the distance between atoms $F–H$ or $O–H$. The calculation was performed by the MPDP/H method; the solvation was taken into account using the point dipole model.

of ions in polar solvents was most often considered as an example. of the following model reaction [69]:

$$FH + NH_3 \rightarrow F^- + NH_4^+,$$

where FH is the acid, and $NH_3$ is the base, therefore in aqueous solutions this system must exist in the ionic form $F^- + NH_4^+$. but in quantum chemical calculations, it was possible to obtain only a local minimum for the ion pair and then only for a fixed and sufficiently long distance between atoms of fluorine and nitrogen. Inclusion of this distance in the number of variable parameters invariably led to the disappearance of a local minimum for the ionized system. As an example Fig. 2.8 shows the results of such a calculation in which the solvation was modelled in the 'supermolecule' approximation by six water molecules.

Upon further investigation of this system [76] it was established that the minimum for an electrically neutral system is local, and the global minimum corresponds to a divided ion pair in which the distance between the ions is about 1 nm (Fig. 2.9). For contact ion pairs, no minimum was found, which indicates the impossibility of their existence in water solutions.

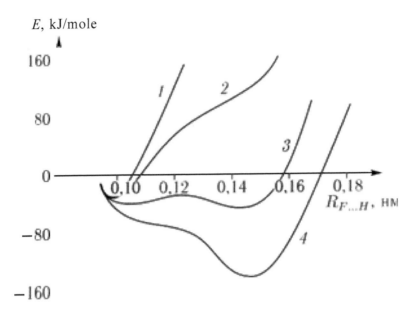

**Fig. 2.10.** PPE cross section for proton transfer reaction. The calculation is completed by the MPDP/N method, a point model was used to take into account solvation dipoles m ($10^{30} \cdot D/m^3$): $1 - 0$; $2 - 0.06$; $3 - 0.12$; $4 - 0.15$.

From these data, the reason for the failure of many authors to consider the reactions of dissociation of molecules in an aqueous medium becomes obvious.

Attempts were made to find contact ion pairs. The model was chosen such that the formation of separated ion pairs was impossible As a result, it was possible to obtain only one minimum for an electrically neutral system.

We are used to believing that with an increase in the polarity of the medium on the profile of the potential energy of the reaction a minimum is formed for the contact ion pair and only then for a divided one. In reality, this is not so. If the solvent can be regarded as a continuous (continuous) medium, then the minima (both local and global) on the potential energy profile for contact ion pairs will be absent at any of its polarity. Really the contact ion pairs in aqueous solutions practically do not form, since the water molecules almost continuously increase the size of the solvation shell with increasing tension of the electric field of the dissolved compound. This continuity effect does not allow contact ion pairs to form.

Contact ion pairs can exist only in solvents that have a qualitatively different structure compared to water and in which the solvation shell cannot continuously increase its size. For example, in acetonitrile, dimethyl sulphoxide and many other polar solvents, the dipole moment is localized at one end of the molecule, and at the other is a nonpolar group of a large size. These solvents are characterized by discrete properties: after filling the first sphere of the solvation shell (the first solvation layer), a sharp decrease in the value of the sequential solvation energies and a second solvation shell practically does not formed.

In the model of point dipoles, the discrete properties of media of this type can be introduced, for example, by limiting the number of point dipoles that can interact with a dissolved molecule.

Figure 2.10 shows the results of calculating the cross section of the potential energy surface for the proton transfer reaction under the assumption that the number of point dipoles of volume $V_{p.d.} = 10^{-3}$ nm$^3$ cannot exceed 200. It can be seen that with this limitation of the number of point dipoles, the global minimum corresponds to contact ion pairs. It is not possible to get it without introducing the discrete element into the used solvation model.

When studying the mechanisms of many organic reactions by quantum chemical methods, it is necessary to take into account the interaction of the reagents and the solvents. This statement primarily refers to ion–molecular reactions. Moreover, the results of quantum-chemical calculations for gas-phase models cannot be used to consider these reactions in solutions, since solvation qualitatively changes the calculated potential energy profile.

To account for solvation, the most widely used methods are currently the Monte Carlo method with atomic–atomic potentials, the 'supermolecule' approximation and the model of point dipoles. All these methods qualitatively correctly convey the change of the profile of the potential energy of ion–molecular reactions during the transition from the gas phase into the solution. The Monte Carlo method gives, apparently, the most reliable results, but its use for calculating potential profiles of organic reactions in polar solvents requires very high costs of computer time and for real reactions is almost impossible.

The 'supermolecule' approximation is used to study the effect of the solvation of the reactivity of organic compounds only in the case of solvents whose molecules contain a small number of atoms.

Such a solvent, in particular, is water. The addition of several water molecules to the reagents makes it possible to understand how a polar solvent affects the potential energy profile of an organic reaction.

The point dipole model can also be used to studying the effect of solvation effects on the reactivity of organic compounds. It gives less accurate results, but is quite suitable for studying solvation effects. at a high level. This model is extremely economical, therefore when solving most applied problems, preference is given to this method. More complex models should only be used. in exceptional cases.

## 2.5. Solvated ion sizes and solution properties

Studies of the effects of electric and magnetic fields on liquid dielectrics have been going on for many decades. In some of them the authors conclude that the action of the field causes dielectric fluid structuring. Studies of the behaviour of a liquid dielectric subjected to a transverse electric field showed that the viscosity of the dielectric varies. This phenomenon, called the 'electroviscous effect', has been described by different authors [77–79]. But the poor reproducibility of the data and often apparent contradictions in the data of different authors did not allow us to clearly distinguish the main mechanism among many phenomena, which may cause viscosity changes. These are: the transmission of angular momentum by ion electrophoresis, the orientation of polar molecules [80], the formation of clusters near the electrode [81], the effect of space charge and electrohydrodynamic effect [82]. All authors, despite different interpretation of the phenomenon, agreed with the need for leakage of electric current through a liquid.

The reason that the effect of electric fields can lead to tangible changes in viscosity is the data on the effect of electric fields on thermal conductivity [83], since this and viscosity, is a characteristic of molecular transport in a liquid.

As a result of the experiment [84], the dependences of the change in viscosity on the applied voltage were obtained as for a cell with one insulated electrode, and for cells with non-insulated electrodes. The changes in the viscosity of the liquid are divided into two groups: in one (polar liquids) a change in viscosity in both cell types (uninsulated and insulated electrodes) occurred in the range of 10–160% with uninsulated electrodes and 5–40% with one insulated electrode. In another group (nonpolar liquids) the change in viscosity did not exceed 6% in non-insulated and 4% in the system of insulated electrodes.

To estimate the interaction energy of the molecules of the studied liquids within the first coordination sphere, we use the option of the London–Debye–Keesom potential [85]

$$\overline{\varphi}_1 - \overline{Z}_1\overline{\varphi}_p = \overline{Z}_1\left[\frac{4}{3}J\alpha^2 + 2\mu^2\alpha + 2\frac{\mu^4}{3KT}\right]\frac{1}{\overline{r}_1^6},$$

where $J$ is the ionization potential, $\alpha$ is the polarizability of the molecule, $\mu$ is the dipole moment of the molecule, $\overline{\varphi}_p$ is the average energy of pairwise intermolecular interaction, $\overline{Z}_1$ is the number averaged over the entire volume of the molecules, $\overline{r}_1$ is the radius of the first coordination sphere.

Studies have shown that when applying an electric field the viscosity changes across the fluid dielectric stream both in polar and nonpolar liquids. Moreover, the effect is realized both when current flows through a liquid and without it.

In the presence of current, the effect is stronger. In this case, we can assume that the momentum transfer is carried out by joint convective and molecular transfer. The main change in viscosity occurs due to an increase in the number of molecules for which a correlation occurs, i.e. due to the formation of new structures in the fluid.

In experimental studies, a phenomenon was discovered consisting in the fact that under the action of an external periodic electric field, solvated ions (clusters) drift [86]. Oriented drift of polarized solvated ions allows one to organize the technological process of elemental enrichment of the aqueous solutions of salts [87]. Thus, the phenomenon of electrically induced selective drift of solvated ions in solutions has a technological application. Further experimental studies have shown that this is not the only application of the discovered phenomenon. In particular, metal salt solutions in polar, liquid dielectrics (water, alcohols, ketones) show interesting properties under the influence of magnetic fields.

The solvated ion is a supramolecular formation – a cluster formed by dielectric molecules associated around the ion. As theoretical and experimental studies [88–90], the cluster size is the value that determines most of the physical and technological properties of solutions. This allows, based on the known physical properties, to evaluate the values of the quantities characterizing individual clusters, for example, experimental results presented in [91].

The relationship between the effective radius of the ion and its limiting molar conductivity is usually described by the equation derived from the Stokes law [92]:

$$r_s = |z_i| eF / 6\pi\eta\lambda_i^0,$$  (2.14)

which includes solvent viscosity $\eta$ and ion charge $z_i$. Equation (2.14) expressing the feedback between the radius of the migrating ion and its mobility and resulting from the motion model of the spherical particle in a continuous medium under the influence of an electric field, is often interpreted as a theoretical justification of the Valden rule $\lambda^0\eta = $ const, which, however, is only observed in particular cases. In the general case, it is necessary to take into account the change in $r_s$ with the degree of solvation, as well as the dependence of $\lambda^0\eta$ on the dielectric permeability and molar volume of the solvent [93, 94]. Besides, moreover, the moving ion, in addition to purely hydrodynamic friction, experiences dielectric braking, which was taken into account in the Zwanzig model [95]

$$\lambda_i^0 = |z_i| eF / \zeta, \quad \zeta = \zeta_V + \zeta_D.$$  (2.15)

Here $\zeta_V$ and $\zeta_D$ are the coefficients of hydrodynamic and dielectric friction, respectively

$$\zeta_V = 6\pi\eta r_i, \quad \zeta_D = \frac{3}{8}\frac{z_i^2 e^2}{r_i^3}\frac{\varepsilon_0 - \varepsilon_\infty}{\varepsilon_0(2\varepsilon_0 + 1)}\tau,$$  (2.16)

where $\varepsilon_0$ and $\varepsilon_\infty$ are the dielectric constant of the solvent at zero and infinite frequencies, $\tau$ is the relaxation time for solvent molecules. There are two assumptions about the motion of an ion in a solvent medium – adhesion of solvent molecules and free ion glide. The friction coefficients in these cases differ in the numerical values (6 and 4 for hydrodynamic, 3/8 and 3/4 for dielectric friction). Most often in the scientific literature, the coefficients of equation (2.16) are used.

Ions with small radii experience small hydrodynamic friction, but at the same time, intensively interact with nearby solvent molecules and experience significant dielectric friction. For large ions, it is clear that hydrodynamic friction comes first. Thus, ions with two different radii can have the same mobility.

The best agreement with experimental data is observed in the Hubbard–Onsager theory [96]. The equations of this theory, are consistent with the equations of the Zwanzig theory. The differences are in dielectric friction coefficients

for sticking and

$$\zeta_D = \frac{17}{280} \frac{z_i^2 e^2}{r^3} \frac{\varepsilon_0 - \varepsilon_\infty}{\varepsilon_0^2} \tau, \tag{2.17}$$

$$\zeta_D = \frac{1}{15} \frac{z_i^2 e^2}{r^3} \frac{\varepsilon_0 - \varepsilon_\infty}{\varepsilon_0^2} \tau, \tag{2.18}$$

for free sliding.

The mobilities of alkali metal cations are much smaller than follows from the Stokes equation (2.14). It indicates noticeable solvation of these ions, as a result of which their effective radii significantly differ from the crystallographic radii $r_{cr}$, since the ions migrate together with the solvate shells. By the values of the effective radii of ions, one can judge the values their solvation numbers.

As follows from Table 2.6, the differences in numerical values for different theories are small. Thus, the Zwanzig and Hubbard–Onsager theories do not have any advantages over the Stokes equations for calculating the effective radii of ions in solutions. At present, there are modifications of the Hubbard–Onsager theory [99–102]. However, the proposed modifications for not introduce significant changes to the description of ionic mobility.

As noted by Robinson and Stokes [103], for large tetra alkylammonium ions the radii of the ions in solution calculated by equation (2.14) can be smaller than crystallographic, which

**Table 2.6.** Characteristics of alkali metal cations in propylene carbonate at 298.15 K

| Cation | $\lambda_i^0 \times 10$ | $r_{cr} \times 10$ | $r_S \times 10$ | $r_Z \times 10$ | $r_{H-O} \times 10$ | $r_S^{corr} \times 10$ | $n_S$ | $n_S$ [98] |
|--------|------|------|------|------|------|------|------|------|
| Li⁺ | 7.2 | 0.60 | 4.65 | 4.58 | 4.63 | 5.77 | 5.9 | 5.8 |
| Na⁺ | 9.68 | 0.95 | 3.37 | 3.18 | 3.32 | 4.73 | 3.3 | 5.1 |
| K⁺ | 11.27 | 1.33 | 2.90 | 2.50 | 2.80 | 4.34 | 2.5 | 4.3 |
| Rb⁺ | 12.31 | 1.48 | 2.65 | – | 2.52 | 4.14 | 2.1 | 3.6 |
| Cs⁺ | 13.07 | 1.69 | 2.5 | – | 4.34 | 4.02 | 1.9 | 3.1 |

Comment. – mobility, $ohm^{-1}m^2$; $r_{cr}$ is the crystallographic radius, nm; $r_S$, $r_Z$, $r_{H-O}$ are effective radii according to the Stokes, Zwanzig and Hubbard–Onsager equations,nm; $r_S^{corr}$ is the corrected radius, nm; $n_S$ is the solvation number.

contradicts the physical meaning. To eliminate this contradiction in [103], and Nightingale in [97] proposed methods of correcting the Stokes radii based on the assumption on the equality of the radii of large tetra alkylammonium ions in solution of their crystallographic radii.

Using the corrected Stokes radii [67–97], the numbers of solvation of cations in propylene carbonate were calculated

$$n_S = \frac{1}{V} \frac{4\pi}{3} \left( r_S^3 - r_{cr}^3 \right), \qquad (2.19)$$

where $V$ is the volume of the solvent molecule, calculated from the molar volume. These values are given in Table 2.6. The table presents for comparison the solvation number of alkali metal ions from Marcus' study [104], obtained from the entropy of solvation ions. Given that the solvation numbers of ions found using different methods can vary quite significantly, sometimes even by an order of magnitude [105], coincidence of the tendency for changes in the numbers of solvation with $r_{cr}$, as well as their absolute values with data [104], can be considered satisfactory.

It can be stated that today there are three approaches to solving the problem of determining the properties of solutions. In a sense, they are all similar, and their main difference is the specifics of determining the dimensions of supramolecular formations – clusters, which are a combination of water and ion molecules around which these molecules are associated.

The fundamental possibility of using the plasma concept of the state of electrolyte solutions was noted in the works of M.M. Baldanov [106, 107]. According to this concept, electrolyte solutions, which are a charge system, can be considered in the plasma-like approximation, since they, as the main structural units of matter, are the basis of the plasma state of matter. The combination of ion properties such as charges and their radii, ionization potentials, energy characteristics and many others, determines the specifics of their behaviour as in solutions, both in gaseous and solid states [108]. Thus, solutions in their main parameters occupy a position close to a low-temperature plasma, characterized by temperature, degree of ionization, density, vibration frequency, and screening parameter.

The basis for calculating the sizes of solvated ions in the plasma-hydrodynamic theory of electrolyte solutions is the Coulomb distributed charge system potential

$$\varphi = \int \frac{\rho dV}{R},$$

(2.20)

where $\rho = e \cdot n$ is the charge density; $e$ is the elementary charge; $n$ is the particle density in 1 cm³ of volume; $R$ is the distance from the system of charges in the elementary volume $dV$ to the observation point. As elementary volume $dV$ we take the volume of a solvated ion equal to $dV = 4\pi r_s^2 \, dr_s$. In this case, the observation point can be taken on the surface of the volume $dV$, then $R$ is equal to the radius of the solvated ion $r_s$. The integration of equation (2.20) under these conditions leads to the expression determining the energy of the interaction of the ion with $n_S$ solvent molecules in the form of:

$$e\varphi = \frac{m\omega^2 r_S^2}{2},$$

(2.21)

where $\omega = \left( \dfrac{4\pi e^2 n}{m} \right)^{1/2}$ is the frequency of Langmuir plasma oscillations.

If the hydrated ion is considered as a system of point charges, then the energy in the form of equation (2.21) can be represented in the shape of

$$e\varphi = \frac{1}{2} \sum \frac{e_a e_w}{R_{aw}} = \frac{1}{2} \frac{zep}{r_S l} n_S,$$

(2.22)

where $e_a = z \cdot e$ is the ion charge; $e_w = p/l$ is the dipole charge of a water molecule; $p$ is the dipole moment of a water molecule; $l$ is the dipole distance; $R_{aw}$ is the radius of the hydrated ion.

Therefore, the equality of representations (2.21) and (2.22) leads to the following expression:

$$\omega^2 = \frac{zep n_S}{m \bar{l} r_S^3},$$

(2.23)

where $n_S$ is the number of solvent molecules surrounding the ion.

Then the corresponding energy of these oscillations of the charge density relative to its equilibrium value is equal to

$$\hbar\omega = \left[ \frac{zep n_S \hbar^2}{m \bar{l} r_S^3} \right]^{1/2}.$$

(2.24)

The boundary of the hydrate complex is determined by the condition $\hbar\omega = k_B T$, where the right side is the kinetic energy of water molecules.

Consequently, the expression for theoretical estimates of the radii of hydrated ions takes the form

$$r_S^i = \left[ \frac{z e p n_S \hbar^2}{m \bar{l} k_B T^2} \right]^{1/2} .$$

(2.25)

Substituting into this equation the values of all universal constants and the water parameters $p$, $T$ in the CGS unit system, we obtain the calculated expression [108]

$$r_S^i = 49.79 \cdot 10^{-8} \left( \frac{z \cdot n_S}{T^2} \right)^{1/3} \text{ cm,}$$

(2.26)

where $n_S = \dfrac{z_i e \bar{l}^2}{r_i p} - \dfrac{3}{2} \dfrac{k_B T \varepsilon \bar{l}^2}{e p}$ is the hydration number in which $r_i$ is

the radius of the ion, $\varepsilon$ is the dielectric constant of the medium.

The values of the radii of solvated ions obtained in the framework of this plasma–hydrodynamic theory of electrolyte solutions, were used to develop a model for assessing the thermal conductivity of aqueous solutions of electrolyte mixtures, as well as for calculating the equivalent electrical conductivity, dynamic viscosity, and coefficients of diffusion of the agents over a wide range of concentrations and temperatures.

The second approach involves multiscale modelling of solvation processes on the basis of a unified (quantum-classical) theory of functional density (UTFD) [109, 110].

This theory is described in the work of M.V. Fedorov, whose goal consisted in the development of general theoretical methods for multiscale modelling of solvation processes and the study of the effect of physicochemical properties of polar solvent on the structural and thermodynamic properties of various solvated objects.

The main attention in the work was given to aqueous solutions as the most important for molecular medicine, biotechnology, food and the drug industry [111].

In this paper, based on the methods of integral equations, a new highly effective method for calculating solvation effects for objects of various sizes, ranging from angstroms to several nanometers. The

**Table 2.7.** Radii of hydrated ions calculated by equation (2.26)

| Ion | $K^+$ | $Rb^+$ | $Be^{2+}$ | $Mg^{2+}$ | $Ca^{2+}$ | $Sr^{2+}$ | $Ba^{2+}$ | $Zn^{2+}$ | $Al^{3+}$ | $Cr^-$ | $Br^-$ |
|---|---|---|---|---|---|---|---|---|---|---|---|
| $r_s \cdot 10^8$, cm | 1.9 | 1.53 | 4.13 | 3.25 | 2.87 | 2.82 | 2.56 | 3.20 | 4.66 | 1.33 | 1.28 |

need for development is due to the disadvantages of the standard theory of functions density (TFD) method in the classical (i.e., not quantum) representation which allows one to determine only the qualitative behaviour of the solution, and the estimates obtained on the basis of this method do not allow accurate calculation of the aggregation energy of complexes.

The UTFD method overcomes the limitations associated with the complexity of direct modelling of the interaction of a salt with solvent molecules, and to study in detail on a molecular scale different structural and energy processes occurring upon dissolution of charged and neutral atoms and molecules in a polar liquid.

The main difference between this method is that it clearly takes into account the structure of the solution, while in other models the structure of the solution is not considered, and the effect of the solvent on the development of complexes is taken into account only through the parameters of interaction.

Formula (2.27) characterizes the main idea of the method – the combination in one functional of the quantum and classical parts of the description systems that are characterized by the electron density $n_e(r)$ and the classical atomic density $n_S(r)$, respectively:

$$F_{\text{total}}\left[n_e(r), n_S(r)\right] = E\left[n_e(r)\right] +$$
$$+ \sum_S \iint n_e(r) u_{eS}(r - R) n_s(r) dr\, dR + F_S\left[n_S(r)\right]. \qquad (2.27)$$

Minimization of this functional with respect to $n_e(r)$ leads to a widely known Kohn–Sham equations, and minimization of this functional with respect to $n_S(r)$ reduces to the problem of solving a system of integral equations similar to the integral equations of fluid theory.

Thus, minimization of the full functional reduces to a self-consistent solution of these equations. Until recently, this was a very non-trivial task due to different characteristic scales of electronic and classical density. This fact explains the not very high interest of

**Table 2.8.** Calculations of the free energy $F$ and average radius $r$ for a solvated electron (index $p$) and a bipolaron (index $b$) [111]

|  | $F_p$ (eV) | $F_b$ (eV) | $r_p$ (Å) | $r_b$ (Å) |
|---|---|---|---|---|
| Water | −1.01 | −1.20 | 2.3 | 2.5 |
| Ammonia | −0.72 | −1.58 | 2.9 | 3.5 |

the scientific community in the use of DFT to describe the quantum classical systems. However, the multiscale wavelet representation of all density functions radically changes situation – wavelet-based minimization algorithms are both stable and computationally efficient [111].

This apparatus was used to develop a number of numerical and theoretical methods by which various phenomena occurring during the solvation of quantum particles (polaron and bipolaron), atomic and molecular ions, polymers, fluorocarbons, polypeptides and their aggregates were investigated. The developed methods allow without compromising the accuracy of the calculation, drastically reduce the time of calculations by comparing to direct numerical simulation using the Monte Carlo and molecular dynamics methods.

Table 2.8 presents the results of the calculation of several thermodynamic and structural parameters for the polaron and bipolaron in water and ammonia.

The third approach was described in [112], where the sizes of supramolecular formations were calculated in the approximation of a 'single ion' from the condition that molecular bridges can form in the electric field of the central ion due to the dipole–dipole interaction of water molecules.

In accordance with the modern concepts of solutions in the vicinity of the solvated ion, molecule, associate, complex, or other similar particle, the structure of the solvent changes in moving away from the centre of the solvated particle. This experimentally confirmed position is reflected in the fact that the solvent molecules of the near and partially far and distant surroundings of the solvated particle are distinguished. For solutions of the electrolytes the concept of 'the boundary of complete solvation' was introduced, very important for elucidating the structure of concentrated solutions of electrolytes. Upon reaching the boundary of complete solvation, all the molecules of the solvents are distributed between the solvate shells of ions, which from this moment 'fight' is conducted for a solvent whose molecules are redistributed depending on the solvation ability of the ions.

From the first principles it follows that the number of solvent molecules associated around the ion is proportional to the tension of the electric field created by this ion. In turn, the field strength is proportional to the charge. It is important that the charge of ions very different in size can be the same. For example, monovalent ions $Cs^+$ and $Li^+$ have an identical charge, but the size of the cesium ion is much larger than the size of the lithium ion.

If we consider an ion as a ball with a charge distributed in its volume, it should be assumed that this charge will be dispersed at the surface of such a ball. This requires the fact that any stable system tends to configuration (state) with a minimum value of potential energy. The ball potential energy with charge $Q$ evenly distributed over its volume is determined by the well-known formula

$$U = \frac{3}{5} \frac{Q^2}{4\pi\varepsilon_0 a},$$

where $a$ is the ball radius, $\varepsilon_0$ is the dielectric constant of the vacuum. The potential energy of a thin spherical layer of the same radius, in which the same charge is dispersed, is determined by the formula

$$U = \frac{1}{2} \frac{Q^2}{4\pi\varepsilon_0 a}.$$

It is seen that the potential energy of the system in the distribution of this charge at the surface will be 5/6 (83.3%) of the potential energy of the system when the charge is distributed over its volume. Thus, the surface charge distribution is less energy intensive.

The field strength of a charge distributed in different spheres with radii $a$ and $b$ are inversely proportional to their radii:
$$E_a/E_b = b/a.$$

Thus, the field strength of the $Li^+$ ion will significantly exceed the field strength of the $Cs^+$ ion, which leads to the fact that the number of solvent molecules associated around the $Li^+$ ion will be noticeably gretare than the number of the molecules around the $Cs^+$ ion. The size of the solvation shell of the lithium ion must also exceed the size of the shell of the cesium ion.

According to the Debye–Hückel theory, the main characteristics that determine the size of a supramolecular formation – a 'cluster' – are the electric potential and its distribution around the ion (cation or anion) located in a continuous medium formed by polar solvent molecules.

It is assumed that the charge distribution around the ion is expressed by the Poisson equation relating charge density to potential:

$$\nabla^2\psi(r) = -4\pi\rho(r)/\varepsilon, \qquad (2.28)$$

where $\nabla^2$ is the Laplace operator; $\psi$ is the potential; $\rho$ is the charge density at the point $r$ for which the potential is calculated; $\varepsilon$ is the dielectric permeability of the medium.

The charge density $\rho$ in the Poisson equation (2.28) is the difference between the number of positive and negative charges per unit volume located at a distance $r$ from the central ion. The potential at this distance is $\psi$.

For salt concentrations in solutions in polar dielectric fluids ensuring the fulfillment of condition $n_i \ll n_{solv}$ where $n_{solv}$ is the number of solvent molecules per unit volume, $n_i$ is the same for the number of ions, we can assume that: the *potential becomes negligible at distances shorter than the distance between most closely spaced ions.*

Then the charge density $\rho$ is nothing but the the polarization charge formed by the polarization of the solvent molecules. In the formalism used in continuous electrodynamics media, it is not required to consider each point charge. The polarization charge is considered continuous, which is the result of statistical averaging.

Solvent molecule polarization can cause appearance in it of both negative and positive polarization charge. If we consider a cation, then its field deforms the generalized electronic shell of the solvent molecule and causes formation of a positive polarization charge at a remote from the cation end of the dipole. If the anion – negative.

Each cation (anion) causes polarization of the surrounding solvent (for example, water): around each cation (anion) an 'atmosphere' is formed with an excess of polarized solvent molecules, screening field of the cation (anion). The generalized electron shell of each polarized solvent molecule is deformed relative to the unperturbed configuration when the total spin molecule is zero. The perturbation is caused by the action of the electric field of the cation (anion). Shell deformation leads to the situation in which that part of the charge of the nuclei or electrons that make up the solvent molecule will be uncompensated in a certain part of that area of space occupied by an undisturbed electron shell. This uncompensated portion of the charge represents a polarization charge of a solvent molecule in an inhomogeneous electric field of a charged particle – a cation (anion). The polarization charge of solvent molecules is determined by the charge of the particles that they shield. The sum of polarization charges of all molecules associated around one particle is equal to the charge of the particle (in absolute terms). Also the polarized charge is positive if the solvent molecules shield the anion, and negative if they shield the cation. This 'atmosphere' ('coat') with an excess of polarized molecules of the solvent also represents a solvate shell.

In any case, the number of polarized molecules of the solvent, having potential $\psi$, is determined by the Boltzmann equation $n_{field} = \bar{n} \cdot \exp\left(-\dfrac{U}{kT}\right)$ where $\bar{n}$ is the average number of solvent molecules per unit solution volume, $U$ is their energy corresponding to the potential $\psi$, $T$ is temperature, $k$ is the Boltzmann constant.

The energy of polarized solvent molecules having the potential $\psi$, is determined by the expression

$$U = q_{H_2O} \cdot \psi, \tag{2.29}$$

where $q_{H_2O}$ is the polarization charge of the molecule. The polarized charge caused by the displacement of the electron distribution induces an additional dipole moment.

Thus,

$$n_{field} = \bar{n} \cdot \exp\left(-\frac{q_{H_2O} \cdot \psi}{kT}\right). \tag{2.30}$$

It is natural to assume that the polarization charge of the molecule (dipole) is proportional to the charge of the ion (cation or anion). Its value is also determined by the number of solvent molecules that 'line up' on the line connecting the ion and the molecule, and shield the ion field. Thus the polarized charge is inversely proportional to the number of solvent molecules that are between the ion and the molecule in question. This is amount is directly proportional to the volume of the sphere that is formed by these 'screening' molecules, that is, the cube of the distance between the ion and molecule. In this regard, the following equation can be written:

$$q_{H_2O} \sim \frac{C \cdot q}{f^3},$$ (2.31)

where $C$ (constant) is the coefficient of proportionality, $q$ is the charge of the ion, $r$ is the distance between the ion and the molecule in question.

Thus, the density of the polarization charge will be determined by the relation

$$\rho = q_{H_2O} \cdot n_{field} = \bar{n} \cdot \frac{C \cdot q}{r^3} \cdot \exp\left(-\frac{C \cdot q \cdot \psi}{r^3 \cdot k \cdot T}\right).$$ (2.32)

The expression defining the value of the constant C depending on the dielectric properties of the solvent, can hardly be just obtained from general physical considerations. But we can say that the equation determining the distribution of potential around the ion has the form

$$\nabla^2 \psi = -\text{const} \cdot q \cdot f(T) \cdot \frac{1}{r^3}.$$ (2.33)

It makes sense to define a relationship that describes the distribution of potential around an ion placed in a dielectric medium, using the equations of electrostatics.

The density of the polarization charge is determined by the ratio

$$\rho_{pol} = -\nabla P,$$ (2.34)

in which the polarization vector $\mathbf{P}$ is linearly related to the electric field vector $\mathbf{E}$ created by the ion:

$$\mathbf{P} = \chi \cdot \varepsilon_0 \cdot \mathbf{E},$$ (2.35)

where $\chi$ is the dielectric susceptibility of the dielectric, $\varepsilon_0$ is the electric vacuum constant.

An ion having a charge $q$ creates a field in a dielectric medium whose distribution is described by the relation

$$\mathbf{E} = \frac{q \cdot r}{4 \cdot \pi \cdot \varepsilon_0 \cdot \aleph \cdot r^3},$$ (2.36)

where $\aleph = 1 + \chi$ is the relative dielectric constant, $\mathbf{r}$ is the radius vector whose origin coincides with the geometric centete of the ion.

Relation (2.34) is a mathematical model of the physical process of the appearance of the resulting charge inside the dielectric caused by inhomogeneous polarization.

Given that the radial component of the vector of electric field strength $E_r = \dfrac{q}{4\cdot\pi\cdot\varepsilon_0\cdot\aleph\cdot r^2}$, and the field of the central ion is spherically symmetric, it is easy to obtain the relation

$$\rho_{pol} = \chi\cdot\frac{q}{2\cdot\pi\cdot\aleph}\cdot\frac{1}{r^3}. \qquad (2.37)$$

As in the previously obtained relation (2.32), the density of the polarization charge is inversely proportional to the third power of the distance counted from the centre of the ion.

In the absence of an electric field, individual solvent molecules are randomly oriented in different directions, therefore, the total dipole moment per unit volume is zero.

In the electric field of the central ion (cation or anion) two processes occur immediately: first, an additional dipole moment is induced due to the forces acting on the electrons (electronic polarizability); secondly, the electric field tends to orient (build) individual molecules, creating the resulting non-zero moment in unit volume. Molecular collisions with their chaotic Brownian motion do not allow them to observe a strictly fixed orientation, but the formation of a supramolecular structure (cluster) is still happening.

Substances that do not conduct electric current are called dielectrics. At not too high temperatures and in conditions when dielectrics are not exposed to very strong electric fields, these substances, unlike conductors, do not contain free carriers of electric current.

Dielectric molecules are electrically neutral and contain an equal number of positive and negative charges. Nonetheless, the molecules have electrical properties. In a first approximation, a dielectric molecule can be considered as a dipole having a dipole electric moment

$$\mathbf{p}_0 = Q\mathbf{l}, \qquad (2.38)$$

where $Q$ is the absolute value of the total positive and total negative charges located at the centers of gravity of these charges, respectively; $l$ is the distance between centers of gravity of the positive and negative charges.

A dielectric is called nonpolar (nonpolar dielectric) if the electrons of the atoms in its molecules are located symmetrically with respect to the nuclei ($H_2$, $O_2$, $CCl_4$, etc.). In such molecules, the centres of gravity of the positive and negative charges coincide in the absence of the external electric field ($l = 0$), and the dipole moment of the molecule is zero. If a nonpolar dielectric is placed in an external electric field, then the deformation of the electron shells in the atoms (molecules) and the centres of gravity of the positive and negative charges are displaced relative to each other. An induced dipole arises in a molecule (atom) of a dielectric proportional to the electric field strength $E$:

$$\mathbf{p}_0 = \varepsilon_0 \alpha \mathbf{E} \ \text{(in SI)},$$
$$\mathbf{p}_0 = \alpha \mathbf{E} \ \text{(in CGSE)}, \qquad (2.39)$$

where $\alpha$ is the polarizability coefficient (polarizability) of the molecule (atom).

The polarizability of a molecule depends only on the volume of the molecule. It is essential that $\alpha$ is independent of temperature. Thermal motion of the molecules of nonpolar dielectrics does not affect the occurrence of the induced dipole moments. Molecules with such dipole moments are similar to quasielastic (induced) dipoles.

A polar dielectric is called such a dielectric the molecules (atoms) of which have electrons located asymmetrically relative to their nuclei ($H_2O$, HCl, $NH_3$, $CH_3Cl$, etc.). In such molecules, the centres of gravity of positive and negative charges do not coincide, being almost at a constant distance $l$ each from other. The molecules of polar dielectrics are similar in their electrical properties to rigid dipoles, for which there is a constant dipole moment

$$\mathbf{p}_0 = \text{const.}$$

A rigid dipole placed in a uniform external electrostatic field experiences the action of a pair of forces with a moment equal to

$$\mathbf{M} = [\mathbf{p}_0 \mathbf{E}]. \qquad (2.40)$$

The moment of the pair $\mathbf{M}$ is directed perpendicular to the plane passing through the vectors $\mathbf{p}_0$ and $\mathbf{E}$, and from the end of $\mathbf{M}$ the rotation goes from $\mathbf{p}_0$ to $\mathbf{E}$ by the shortest anti-clockwise path.

In real molecules of polar dielectrics, in addition to rotation of the axes of the dipoles along the field, molecules are deformed, and some induced dipole moment is created in them.

If the polar dielectric is not in an external electrical field, then as a result of chaotic thermal motion of molecules the vectors of their dipole moments are randomly oriented. Therefore, in any physically infinitesimal volume $\Delta V$ the sum of dipole moments of all molecules is zero. (Moreover, $\Delta V \gg v_0$, where $v_0$ is the volume of one molecule, and the volume $\Delta V$ contains a very large number of molecules.)

In a nonpolar dielectric that is not in an external electric field, dipole induced moments of molecules cannot form. When a dielectric is introduced into an external electric field, the dielectric becomes polarized, consisting in the fact that in any elementary volume $\Delta V$, the total dipole non-zero moment of molecules appears. The dielectric which is located in this state is called polarized (polarized dielectric). Three types of polarization are distinguished depending on the structure of the molecules (atoms) of the dielectric.

.a) *Orientational polarization in polar dielectrics.* An external electric field tends to orient the dipole moments of hard dipoles along the direction of the electric field. This is prevented by the chaotic thermal motion of molecules, tending randomly to 'scatter' dipoles. As a result of the joint action of the field and thermal motion, there arises a preferential orientation of the dipole electric moments along the field, increasing with increasing electric field strength and with decreasing temperature.

b) *Electronic (deformation) polarization in nonpolardielectrics.* Under the influence of an external electric field, molecules of dielectrics of this type induce dipole moments directed along the field. Thermal motion of molecules does not affect electronic polarization. In gaseous and liquid dielectrics electron polarization also occurs almost simultaneously with the orientational polarization.

c) *Ion polarization in solid dielectrics having ioncrystalline lattices.* For example, NaCl, CsCl, etc. The external electric field causes displacement of all positive ions in the direction of tension $E$, and all negative ions – in the opposite direction.

A quantitative measure of dielectric polarization is the polarization vector **P**. The polarization vector (polarization) is the ratio of the electric dipole moment of the small volume $\Delta V$ of the dielectric to the magnitude of this volume

$$\mathbf{P} = \frac{1}{\Delta V}\sum_{i=1}^{N}\mathbf{p}_i,$$

(2.41)

where $\mathbf{p}_i$ is the electric dipole moment of the $i$-th molecule; $N$ is the total the number of molecules in the volume $\Delta V$.

This volume should be so small that inside it the electric field could be considered homogeneous. At the same time the number $N$ of molecules in the volume $\Delta V$ should be large enough so that statistical research methods can be applied.

For a homogeneous nonpolar dielectric in a uniform electric field,

$$\mathbf{P} = \bar{n}\,\mathbf{p}_0, \tag{2.42}$$

where $\mathbf{p}_0$ is the dipole moment of one molecule..

Using the formula for $\mathbf{p}$, we obtain

$$\mathbf{P} = \bar{n}\varepsilon_0\alpha\mathbf{E} = \varepsilon_0\chi\mathbf{E} \text{ (in SI)},$$
$$\mathbf{P} = \bar{n}\alpha\mathbf{E} = \chi\mathbf{E} \text{ (in the CGSE)}, \tag{2.43}$$

where the dielectric susceptibility of the substance is $\chi = \bar{n}\alpha$.

For a homogeneous polar dielectric located in a homogeneous electric field

$$\mathbf{P} = \bar{n}\langle\mathbf{p}_0\rangle \tag{2.44}$$

where $\mathbf{p}_0$ is the average value of the constant dipole component of the moment of the molecule along the field strength.

If the polar dielectric is in a weak external electric field, then the dielectric susceptibility is calculated by the Debye–Langevin equation:

$$\chi = \frac{\bar{n}p_0^2}{3\varepsilon_0 kT} \text{ (in SI)},$$

$$\chi = \frac{\bar{n}p_0^2}{3kT} \text{ (in CGSE)}. \tag{2.45}$$

Given that $\chi = \aleph - 1$, it is easy to obtain the relation

$$\rho_{\text{pol}} = \frac{\bar{n}\cdot p_0^2}{3\cdot\varepsilon_0\cdot k\cdot T}\cdot\frac{q}{2\cdot\pi\cdot\aleph}\cdot\frac{1}{r^3}, \tag{2.46}$$

and the Poisson equation is written in the form

$$\nabla^2\psi = -\frac{2}{3}\cdot\frac{\bar{n}\cdot p_0^2\cdot q}{(\varepsilon_0 - \aleph)^2\cdot k\cdot T}\cdot\frac{1}{r^3}, \tag{2.47}$$

which is consistent with expression (2.33) obtained in the framework of the Debye–Hückel approximation.

It is convenient to rewrite equation (2.47) in the form

$$\frac{d^2\psi}{dr^2} + \frac{2}{r}\frac{d\psi}{dr} = \frac{a}{r^3},$$ (2.48)

where $a = -\dfrac{2}{3}\cdot\dfrac{\bar{n}\cdot p_0^2\cdot q}{(\varepsilon_0 - \aleph)^2\cdot k\cdot T}$

The boundary conditions for the Poisson equation as applied to the case under consideration are written in the form

$$\psi\,(r = r_0) = \psi_0;\ \psi\,(r \to \infty) = 0,$$ (2.49)

where $r_0$ is the ionic radius, $\psi_0$ is the ion potential. Equality of potential to zero at an infinite distance from the point charge follows from the basic provisions of electrostatics and electrodynamics of continuous media. The potential of an ion is defined as the potential of a uniformly charged sphere (charge $q$) with radius $r_0$:

$$\psi_0 = \frac{1}{4\pi\varepsilon_0}\frac{q}{r_0},$$ (2.50)

for the ion $Y^{3+}$ $q = 3\cdot 1.6\cdot 10^{-19} = 4.8\cdot 10^{-19}$ C, $r_0 = 1.06\cdot 10^{-10}$ m, $\varepsilon_0 = 8.854\cdot 10^{-12}$ C/V $\cdot$ m. Thus, for the ion $Y^{3+}$ $\psi_0 = 40.699$ V.

Equation (2.48) is a linear inhomogeneous second-order equation with variable coefficients. Its solution has the form

$$\psi(r) = -\frac{1}{r}\left(\ln\frac{C_1}{r^a} - a\right) + C_2,$$ (2.51)

where $C_1$ and $C_2$ are constants determined by the boundary conditions (2.43). It can be seen that in accordance with the second boundary condition $C_2 = 0$, and the use of the first makes it possible to determine

$$C_1 = r_0^a \exp\,(r_0\,\psi(r_0) + a).$$ (2.52)

For example, in the SI system, $C_1$ is 1 for $Y^{3+}$ ions and $Ce^{3+}$. The value of $a$ in the SI system is $-1.951\cdot 10^{-10}$ for $T = 298$ K and $q = 3|\bar{e}|$, where $\bar{e}$ is the electron charge.

The electric field strength (as a function of $r$), in a spherical coordinate system, in the case of central symmetry is determined by the relationship

$$E_r(r) = -\frac{d\psi}{dr} = -\frac{1}{r^2} \cdot \ln \frac{C_1}{r^a}. \tag{2.53}$$

Theoretical and experimental studies of the formation of nanometer functional structures [113] show the dipole–dipole interaction of water molecules may cause molecular bridges to form in the interelectrode gap. In this case, there is a critical electric field $E_{cr}$ for the formation of molecular bridges:

$$E_{cr} = \frac{1}{\alpha} \left[ \left( p_0^2 + 2 \cdot \alpha \cdot k \cdot T \right)^{1/2} - p_0 \right], \tag{2.54}$$

where $\alpha$ is the polarizability of a water molecule.

Figure 2.11 shows the form of the function $E_r(r)$ for $Li^+, Cd^{2+}, Y^{3+}$ cations within $0.1~\mu m \le r \le 0.75~\mu m$.

For $E > E_{cr}$, the polarized molecules will be bound by a dipole–dipole interaction and oriented in the field direction of the central ion. For $E < E_{cr}$, the thermal motion of the molecules must destroy the bridges. The cluster radius can be estimated from the condition $|E| = |E_{cr}|$, i.e., from the relation

$$\frac{1}{r_{cr}^2} \cdot \ln \frac{C_1}{r_{cr}^a} \approx \frac{1}{\alpha} \left[ \left( p_0^2 + 2 \cdot \alpha \cdot k \cdot T \right)^{1/2} - p_0 \right]. \tag{2.55}$$

This condition means that for $r = r_{cl}$ the field of the central ion is 'balanced' by the opposite direction of the field of 'bridges' of the lined up dipoles. Thus, the 'fur coat' of the lined dipoles shield the central ion.

The solution of the transcendental equation (2.55) gives the values of $r_{cl}$, which are summarized in Table 2.9.

As can be seen from the data presented, the size of the solvated metal ion in case of dissolution of its salt in water at room temperature is about 0.5 μm. Of course, this is a time-averaged size value. The value itself fluctuates relative to the value indicated in the table. One can talk about that ib the intervals exceeding the characteristic time between Brownian collisions, the size of the cluster formed by the ion and the water molecules associated around it indicated in the Table. With increasing temperature of the solvent the cluster size is reduced.

The formation of so-called *ion–dipole clusters* of the type $A^+(AB)_m$, where $AB$ is a polar molecule (including an ion pair), $m$ is the number of polar molecules, is possible also in concentrated solutions of polar liquids (for example, butyl alcohol in transformer oil) [114].

It is believed that such structures can play a major role in the development of EHD flows and in the formation of specific bipolar near-electrode structures [115, 116]. The kinetics of the formation of an ion–dipole cluster is determined by the sequence of reactions

$$A^+ + AB \underset{k_{12}}{\overset{\alpha_{12}}{\rightleftarrows}} A^+(AB),$$

$$A^+(AB) + AB \underset{k_{22}}{\overset{\alpha_{22}}{\rightleftarrows}} A^+(AB)_2.$$

$$\cdots\cdots\cdots\cdots\cdots\cdots\cdots\cdots\cdots\cdots\cdots\cdots\cdots \quad (2.56)$$

$$A^+(AB)_{m-1} + AB \underset{k_{m2}}{\overset{\alpha_{m2}}{\rightleftarrows}} A^+(AB)_m.$$

Here the designation of the reaction constants is changed: $\alpha_{i2}$, $k_{i2}$ are the speeds of recombination and dissociation in the $i$-th reaction.

To obtain a criterion for the existence of ion–dipole clusters, it is assume that the dipoles fill only the first coordination sphere near the $A^+$ ion and the influence of dipoles in the ion–dipole cluster on the processes of dissociation and recombination in the $i$-th reaction can be neglected. In this case, one can obtain [114]

$$\alpha_{n2} = 2.7\pi\left(D_{n-1} + D_2\right)r_d, \quad (2.57),$$

$$k_{n2} = \left(D_{n-1} + D_2\right)r_B^{-2}\delta^{-7/2}\exp\left(-\frac{1}{\delta}\right), \quad \delta = \frac{R_{12}}{r_B}, \quad (2.58)$$

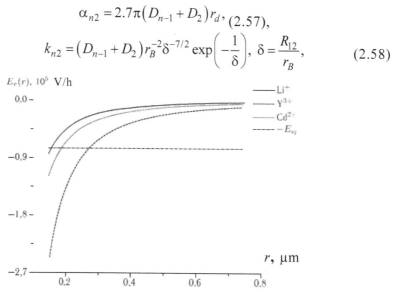

Fig. 2.11. Dependence $E_r(r)$ for Li$^+$, Cd$^{2+}$, Y$^{3+}$ cations for a temperature of 298 K.

**Table 2.9.** Cluster sizes for ions $Y^{3+}$, $Ce^{3+}$, $La^{3+}$, $Sc^{3+}$, $Pr^{3+}$ at temperatures of 298 and 313 K

| Cation | $r_{cl}$, μm | |
|--------|:---:|:---:|
|  | $T = 298$ K | $T = 313$ K |
| $Y^{3+}$ | 0.275478 | 0.270399 |
| $Ce^{3+}$ | 0.275433 | 0.270357 |
| $La^{3+}$ | 0.275127 | 0.270066 |
| $Sc^{3+}$ | 0.276777 | 0.271626 |
| $Pr^{3+}$ | 0.275478 | 0.270499 |

where $R_{12}$ is the minimum distance to which ion $A^+$ approaches the dipole, $r_d = \sqrt{2R_{13}r_B}$ is the effective size of the ion–dipole cluster, $R_{13}$ is the dipole length, $D_{n-1}$, $D_2$ are the diffusion coefficients, $r_B$ is the radius of the ion pair, determined according to the Semenchenko –Bjerrum concept [117, 118] as

$$r_B = \frac{e_1 e_2}{8\pi\varepsilon k_B T}, \qquad (2.59)$$

where $e_1$, $e_2$ are the charges of ions, $\varepsilon$ is the absolute dielectric medium permittivity, $k_B$ is the Boltzmann constant, $T$ is the absolute temperature.

From (2.57) it follows that the association constants of all reactions (2.56) are identical and equal to the association constant of ionic 'tees' $K_{30}$. Here $N$ is the concentration of dipole molecules $AB$, $n$ is the concentration of free ions $A^+$, $n_i$ is the concentration of ion-dipole complexes $A^+(AB)_i$. According to the chemical kinetics the concentration $n_i$ are determined by the following system of equations:

$$\frac{dn_1}{dt} = \alpha_{12} nN - k_{12} n_1,$$

$$\frac{dn_2}{dt} = \alpha_{22} n_1 N - k_{22} n_2,$$

$$\dots\dots\dots\dots\dots\dots\dots\dots \qquad (2.60)$$

$$\frac{dn_m}{dt} = \alpha_{m2} n_{m-1} N - k_{m2} n_m,$$

From (2.58), (2.60) it follows that the characteristic transit time of the $i$-th reactions is

$$\tau_i = \frac{1}{k_{i2}} = \frac{r_d^r}{D_2 + D_{i-1}} \delta^{7/2} \exp\left(\frac{1}{\delta}\right). \qquad (2.61)$$

At typical $R_{12} \sim R_{13} \sim 7$ Å, $r_B \sim 126$ Å, $D_2 \sim 10^{-6}$ cm$^2$ s$^{-1}$ values, the characteristic times $\tau_i$ are extremely small ($\tau_i \sim 10^{-4}$ s); therefore, in experiments, the concentration distribution is established almost instantly and depends only on the degree of mixing of the solution.

In equilibrium from (2.60) we have

$$n_1 = (K_{30}N)n, \ n_i = (K_{30}N)n_{i-1} = (K_{30}N)^i n, \ i \geq 2. \qquad (2.62)$$

This shows that for $K_{30}N < 1$, the equilibrium concentrations of ion-dipole clusters containing $i$ dipoles decrease exponentially, while for $K_{30}N > 1$ sedimentation instability occurs. In this case, ion–dipole complexes evolve into droplet macromolecules, which, sticking together, precipitate. The presence of such sedimentation instability is often observed in experiments, which leads to poor solubility of polar liquids in nonpolar ones.

The distance is crucial for calculating the dissociation rates, and it must be calculated taking into account the interaction of charged components with the environment (solvation). Obviously, in this case, the effective ion size is much larger than its effective radius without considering solvation (radius of free ions). This may be the reason for the difference in the calculation of constants of the reaction rates by statistical methods (for example, by the method of statistical sums [119]) and kinetic methods. At present time this circumstance is realized, for example, local polarization of the medium is the main factor in calculating the mobility of both electrons [120] and ions [121].

| Reaction | Recombination coefficient α | Dissociation rate $k$ | Characteristic size of ionic complexes | Small parameter |
|---|---|---|---|---|
| | | | | |
| | | | | |

In the first two approaches (for obvious reasons), the properties of solutions are determined mainly by the energy of many-particle interactions of ions in a dielectric medium. And this unambiguously assumes the attraction of such individual characteristics as solvation numbers, masses of solvated ions, energy of intermolecular interactions of the solvent, activity coefficients, sizes of solvated ions. According to a series of experiments, their determination is impossible in principle (electroneutrality condition), only theoretical estimates or semi-empirical processing of experimental material based on a number of plausible but essentially arbitrary assumptions are probable.

The validity of the assumptions can be determined experimentally, for example, by comparing the observed and theoretical values of a parameter whose value directly depends on characteristics of the solution, determined in the framework of a particular approach. This kind of work was started back in the 30s of the past.century [122].

The radii of the solvated ions obtained in the plasma-hydrodynamic model of electrolytes or in the simulation of solvation based on CDFT (combined density functional theory) and given in the works are in angstrom units. The same values are obtained in the framework of the classical theory of electrolytes proposed by Samoilov. In so many modern studies they are being specified (for example, in studies of the electrical conductivity of solutions [123]),

Generally speaking, the study of the electrical conductivity of solutions is of great importance for the study of solvation and electric transport properties of ions in the electrolyte solutions. These properties depend on both the charge and radius of the ions, as well as the degree of their hydration [124–126]. In the framework of the Samoilov theory [124], the positive and negative hydration of the ions are determined depending on the mobility of the molecules of water in the hydration shell of the ions compared to pure water.

The movement of hydrated ions in the solution is accompanied by a constant exchange of the surrounding water molecules. Different in nature, the ions are characterized by different activation energies of translation of the water molecules from their hydration shell from one equilibrium provisions to another.

The observed dependence in the change in the electric transport characteristics of ions can be explained by differences in their sizes and hydration numbers. Based on equation (2.63) following from the Stokes law [127–129], one can calculate the Stokes radius of a moving solvated ion

$$r_S = \frac{|z_\pm| F^2}{6\pi N_A \eta \lambda_\pm^0},$$
(2.63)

where $z_\pm$ is the ion charge, $F$ is the Faraday constant, $\lambda_\pm^0$ is electrical conductivity.

To establish the effective radius of the ions, Gill [127, 128] proposes the following empirical equation:

$$r_{ef} = r_S + 0.0103\varepsilon + r_y, \qquad (2.64)$$

where $\varepsilon$ is the dielectric constant of water; $r_y$ is a parameter equal to 0.85 Å for non-associated solvents and 1.13 Å for associated solvents with high values of the dielectric constant $\varepsilon$ (water). Using the found effective ion radius and equation [126, 129]

$$n_S = \frac{4\pi}{3V_L}\left(r_{ef}^3 - r_{cr}^3\right), \qquad (2.65)$$

it is possible to calculate the magnitude of the double number of the ions $n_S$; here $V_L$ is the volume occupied by one water molecule and equal to 12.2 Å [124], $r_{cr}$ is the crystallographic radius of ions [130]. The calculated ion parameters at 298 K are presented in Table 2.10. From the table it can be seen that the anion of selenate is characterized by the largest effective radius and hydration number. The hydration number of the hydrogen cation is equal to 1.89, i.e., approaches 2, since the ion is $H^+$, which is relayed, currently owned by two neighbouring water molecules. The smaller the radius and the lower the positive hydration of ions, the less should be their hydration number. This is typical for the alkali metal ions [101], while polyatomic ions with large radii $(PO_4^{3-}, TeO_4^{2-}, SeO_4^{2-}$ and others)have $n_S$ values close and even larger thn 20 [132]. This is an explanation of their high positive hydration.

The fact that electrical conductivity and, therefore, the size of the solvated ion depends on the concentration of the salts in the solvent is also noted [133].

Among the fifty ions studied [134], selenitic and tellurite different ions are characterized by the highest values of positive hydration, i.e., the mobility of water molecules in their hydration shell and the rate of its rearrangement are minimal, and the structure of the hydration shell is the most stable.

It is known [135, 136] that the maximum at the isotherms of electrical conductivity is a mandatory property of the electrolyte–wtare system. Its presence is explained by the opposite effect of two factors: an increase in the number of ions and a decrease in the mobility of ions due to viscosity increase with increasing solution concentration. From the viewpoint of Samoilov's theory [131], the

**Table 2.10.** Equivalent electrical conductivity at infinite dilution, radii and hydration numbers of ions at 298 K

| Ion | $r_{cr}$, Å | $r_s$, Å | $r_{ef}$, Å | $n_s$ |
|---|---|---|---|---|
| $H^+(H_3O^+)$ | 1.40 | 0.263 | 2.202 | 1.89 |
| $Na^+$ | 0.98 | 1.84 | 3.78 | 18.20 |
| $SeO_4^{2-}$ | 3.05 | 2.45 | 4.39 | 19.30\ |

**Table 2.11.** Effective radius of ions in water at 0 and 25°C

| Ion | Radius $r$, Å |
|---|---|
| $Na^+$ | 0.98 |
| $K^+$ | 1.33 |
| $Cs^+$ | 1.65 |
| $Mg^{2+}$ | 0.78 |
| $SeO_3^{2-}$ | 2.76 |
| $TeO_3^{2-}$ | 2.85 |
| $Hg(CN)_4^{2-}$ | 4.68 |

presence of a concentration maximum in the course of electrical conductivity isotherms is associated with rearrangement of the structure of the solution from water-like to the structure of a liquid crystal of the electrolyte [137–143].

The most common explanation for negative phenomena or positive hydration of ions based on the consideration of the kinetic stability of the supramolecular formation – a cluster, formed by ions and water molecules. Kinetic stability is determined by the activation energy of the process of exchange of water molecules near the ion to water molecules in the solvent mass, i.e., by a change under the action of ions of potential barriers separating neighbouring equilibrium positions of water molecules. The kinetic stability can be judged by the rate of their exchange and the time of binding with the ions. With this approach, hydration is considered not as the binding of one or another number of water molecules in the solution, but as the action of ions on the translational motions of the nearest water molecules [144–146]. Table 2.11 gives the values of the effective radius for various ions obtained by taking into account the exchange of water molecules between the solvate shell and the water surrounding the hydrated ion.

The use of the values of the radii of solvation shells, which are units or even tens of angstroms, in the theoretical explanation of

**Table 2.12.** Cluster sizes and expected characteristic frequencies for $Y^{3+}$, $Ce^{3+}$, $La^{3+}$, $Sc^{3+}$, $Pr^{3+}$ ions at temperatures of 298 and 313 K

| Cation | $T = 298$ K | | $T = 313$ K | |
|---|---|---|---|---|
| | $r_{cl}$, μm | $v_p$, Hz | $r_{cl}$, μm | $v_p$, Hz |
| $Y^{3+}$ | 0.275478 | 1481.59 | 0.270399 | 1537.77 |
| $Ce^{3+}$ | 0.275433 | 942.18 | 0.270357 | 977.89 |
| $La^{3+}$ | 0.275127 | 951.07 | 0.270066 | 987.05 |
| $Sc^{3+}$ | 0.276777 | 2902.82 | 0.271626 | 3013.96 |
| $Pr^{3+}$ | 0.275478 | 935.19 | 0.270499 | 969.93 |

**Table 2.13.** Frequencies for ions $Y^{3+}$, $Ce^{3+}$, $La^{3+}$, $Sc^{3+}$, $Pr^{3+}$ at a temperature of 298 K

| Cation | Solvate shell radius $r_{cl}$, μm | Number of $H_2O$ molecules in solvate shell, $\times10^9$ mol | Cooperative rotational movement of $H_2O$ molecules combined into a solvation shell, relative to an axis passing at a distance $R = r_{cl}$ from the center of inertia, $K_r = 1.4$; kHz | Cooperative rotational movement of $H_2O$ molecules combined into a solvation shell relative to the centre of inertia, $K_r = 0.4$; kHz | Frequency of the rotational motion of the cluster as a whole, kHz | The frequency of the transition of the vibrational motion to the rotational one, kHz; $K_r = 0.4$ |
|---|---|---|---|---|---|---|
| $Y^{3+}$ | 0.275478 | 2.93 | 1.058 | 3.70398 | 1.828 | 1.852 |
| $Ce^{3+}$ | 0.275433 | 2.929 | 0.67298 | 2.35544 | 1.162 | 1.178 |
| $La^{3+}$ | 0.275127 | 2.919 | 0.6793 | 2.37767 | 1.173 | 1.189 |
| $Sc^{3+}$ | 0.276777 | 2.972 | 2.073 | 7.25705 | 3.581 | 3.629 |
| $Pr^{3+}$ | 0.275478 | 2.93 | 0.66799 | 2.3379 | 1.154 | 1.169 |

the phenomenon (effect) of selective drift of solvated cations under the asymmetric electric field leads to estimated values of the field frequencies, at which the manifestations of the effect, amounting to hundreds of megahertz should be expected. Nevertheless, the phenomenon occurs at frequencies not exceeding units of kilohertz, and at certain salt concentrations in water – tens of hertz [6, 8].

For this reason, it became necessary to search for other approximations and assumptions when describing the process of solvation. Besides, there are experimental data indicating that the clusters represent stable formations with characteristic sizes of about 1 μm [147].

Table 2.12 shows the values of the radii of the clusters defined from condition (2.55), and the frequency values corresponding to the main cluster rotation state.

Table 2.13 shows the frequencies corresponding to various components of the rotational–translational motion of cationic aquacomplexes, and the frequencies corresponding to the transition of oscillatory movements to rotational one. The values of the frequencies presented in Table 2.13 were obtained from the relations (2.9)–(2.11) for $J = 1$ (minimum values).

The cluster sizes thus determined and the characteristic values of the frequencies of the electric field made it possible to estimate the values of the parameters at which the manifestation of the effect of electroinduced drift of aquacomplexes in the aqueous solutions of the salts should be expected [148, 149]. Indeed, oriented drift was observed experimentally at frequencies of units of kHz. Further experimental studies of the conditions for the maximum manifestation of the effect showed that in certain cases the total separation coefficient reaches large values at significantly lower frequencies. This necessitated the estimation of the cluster sizes using other assumptions and approximations.

## Conclusions

In accordance with the law on the equal distribution of energy over degrees of freedom, each classical degree of freedom of a molecule in a state of statistical equilibrium has an energy equal to ($kT/2$). Each structureless particle has three degrees of freedom corresponding to three independent motions along the $x$, $y$, and $z$ coordinates. Polyatomic nonlinear solvent molecule, consisting of $N$ atoms, has $3N$ degrees of freedom, of which three are due to the movement of the centre of gravity, that is, the movement of the molecule as a whole, three – by rotating the molecule as a whole. The remaining ($3N-6$) degrees of freedom describe the vibrational motion of atoms in a molecule of the solvent. The solvate shell is formed by $M$ solvent molecules. Accordingly, it has $3M$ degrees of freedom. The solvated cation is formed by $M$ solvent molecules and the cation itself. The metal cation can be considered a structureless particle having 3 degrees of freedom. Thus, the solvated cation has $9N(M + 1)$ degrees of freedom.

In a state of equilibrium, as already indicated, for each degree of freedom there is a very definite energy. Thus, $Z$ solvated cations in a salt solution in a polar dielectric fluid have translational energy equal to $Z(3kT/2)$, rotational motion $Z(3kT/2)$, energy of the oscillations $Z(3N-6)(3M-6)kT$. Quantum effects reduce the stock of energy. In addition to the solvated cations, there are solvated anions and non-associated solvent molecules (solvent). For each degree of freedom of each (each) of them there is a very definite energy.

Figuratively speaking, the energy reserve of a salt solution in a polar dielectric liquid in an equilibrium state is distributed over three energy reservoirs corresponding to translational, rotational and vibrational degrees of freedom (we consider that temperatures are such that electronic degrees of freedom are not excited). These reservoirs communicate, and energy can relatively free overflow from one reservoir to another (energy exchange between various degrees of freedom occurs in collisions of solvated ions (clusters)). Suppose now that we have changed (for example, increased) the energy supply in one of the containers. It is clear that the excess energy will immediately begin to overflow into neighbouring ones. So there is an idea of nonequilibrium (irreversible) processes that occur due to the exchange of energy in collisions. The irreversible processes leading to installation in the system of a statistical equilibrium, are described as relaxation processes.

Nonequilibrium in the solution occurs if the solution is exposed to an external impact. In our case, to the action of an external periodic electric field, the solution being electrically insulated from the electrodes forming this field.

Imbalance also occurs in a weak electric field. It is significant that small perturbations of the equilibrium distribution function (in principle) can determine not the corrections to the effects, but the effects themselves.

One of the main characteristics of the process of establishing equilibrium is the energy relaxation time at a given degree of freedom. This characteristic of the relaxation process sets the time scale (or spatial scale) of the existence of nonequilibrium. If the phenomenon under investigation is considered in a period of time, much longer than the relaxation time, we can assume that the relaxation processes have basically already ended and the state of local equilibrium has been established. If the characteristic time of the studied phenomenon is of the order of relaxation time, the nonequilibrium appears to be significant.

The relaxation time of the energy of an external electric field at the vibrational degrees of freedom of solvated ions (clusters) are incommensurably longer than the relaxation time on the vibrational degrees of freedom of the molecules that form the solution. Studies have shown that it reaches milliseconds. The effect of electrically induced selective drift of solvated ions in salt solutions in polar dielectric liquids occurs with characteristic times and spatial scales commensurate with the corresponding scales of relaxation processes in the solvate shell. This allows counting on the discovery of new significant nonequilibrium effects.

Further development of studies of the processes accompanying the effect of periodic electric and magnetic fields on salt solutions in polar dielectric liquids will contribute to the development of nonequilibrium molecular physics as a science about physical and chemical processes,

# 3

# Self-consistent electric field in the volume of the salt solution

## 3.1. Fluctuations in the polarization charge in volume of the solution; solvated ion sizes and frequencies excitations of electroinduced drift

We assume that each cation (anion) causes polarization of the surrounding solvent (water): around each cation (anion), an 'atmosphere' is formed with an excess of polarized water molecules, screening the field of the cation (anion). The generalized electron shell of each polarized water molecule is deformed relative to the unperturbed configuration when the total spin of the molecule is zero. The disturbance is caused by an electric field. of the cation (anion). The deformation of the shell leads to the fact that part of the charge of the nuclei or electrons that make up the water molecule will be uncompensated in a certain part of the region of space occupied by the unperturbed electron shell. This uncompensated part of the charge is the polarization charge of a water molecule in the inhomogeneous electric field of a charged particle – a cation (anion). In this case, the polarization charge of the water molecules is determined by the charge of the particle that they shield. The sum of the polarization charges of all molecules associated around one particle is equal to the charge carrier (in absolute terms). In addition, the polarization charge is positive if the water molecules screen the anion, and negative if they screen the cation. This 'atmosphere' ('fur coat') with an excess of polarized water molecules is a solvation shell.

Consider the simplest model of electron polarization of a solvent molecule. In this case, a single molecule can be represented as a mechanical oscillator, when a charge having a mass, under the influence of electric forces, carries out a forced oscillatory motion near the centre of mass. In this case, the deviation of the polarizable charges from the equilibrium position is determined by the magnitude of the electric field and the coefficient of elasticity characterizing the elasticity of the binding forces of charges in the molecule. These quantities are related by

$$-\omega^2 \mathbf{r}_m + \frac{\beta}{m}\mathbf{r}_m = \frac{e}{m}\mathbf{E}, \tag{3.1}$$

where $\mathbf{r}_m$, $\beta$ are parameters, the first of which is the deviation of the charges from the equilibrium position, and the second is the elasticity coefficient characterizing the elasticity of the electric forces of the binding of charges in the molecule. We introduce the concept of resonance frequency related
    charges

$$\omega_0 = \frac{\beta}{m}; \tag{3.2}$$

then from (3.1) we can obtain the relation

$$r_m = -\frac{e \cdot E}{m\left(\omega^2 - \omega_0^2\right)}. \tag{3.3}$$

It is seen that in relation (3.3), as a parameter, the frequency of natural vibrations, which includes the mass of the charge, is already present. This suggests that the inertial properties of oscillating charges will affect the vibrational processes of polarized atoms and molecules.

Inertial properties play the role of inductance in the electrical circuit. We call this 'kinetic inductance'. This determines the dependence of the magnitude of the polarization vector on the frequency of the electric field. We introduce the polarization vector

$$\mathbf{P} = -\frac{n \cdot e^2}{m} \cdot \frac{1}{\left(\omega^2 - \omega_0^2\right)}\mathbf{E}. \tag{3.4}$$

The dependence of the polarization vector on the frequency is associated with the presence of mass of links and their inertia does not allow this vector to precisely follow the electric field, reaching its static value at each moment of time. When the field frequency in the dielectric volume coincides with the resonant frequency of an individual molecule the polarization vector tends to infinity. This means there is resonance at this frequency. Since electric induction is determined by the relation

$$\mathbf{D} = \varepsilon_0 \mathbf{E} + \mathbf{P}, \tag{3.5}$$

then the second Maxwell equation takes the form

or

$$\operatorname{rot}\mathbf{H} = j_\Sigma = \varepsilon_0 \frac{\partial \mathbf{E}}{\partial t} + \frac{\partial \mathbf{P}}{\partial t}. \tag{3.6}$$

$$\operatorname{rot}\mathbf{H} = j_\Sigma = \varepsilon_0 \frac{\partial \mathbf{E}}{\partial t} - \frac{n \cdot e^2}{m} \frac{1}{\left(\omega^2 - \omega_0^2\right)} \frac{\partial \mathbf{E}}{\partial t}. \tag{3.7}$$

where $j_\Sigma$ is the total current flowing through the volume of the dielectric.

In this expression, the first term on the right-hand side is the bias current in vacuum, and the second is the current associated with the presence of bound charges in the dielectric molecules. In expression (3.7), the specific kinetic inductance of the charges participating in the vibrational process $L_{kd} = m$ appeared again

$$L_{kd} = \frac{m}{n \cdot e^2}, \tag{3.8}$$

This kinetic inductance determines the kinetic inductance of the bound charges. This is natural, since the oscillating charges also have masses, therefore, have inertia.

With this in mind, relation (3.7) can be rewritten in the form

or

$$\operatorname{rot}\mathbf{H} = j_\Sigma = \varepsilon_0 \frac{\partial \mathbf{E}}{\partial t} - \frac{1}{L_{kd}} \frac{1}{\left(\omega^2 - \omega_0^2\right)} \frac{\partial \mathbf{E}}{\partial t} \tag{3.9}$$

$$\operatorname{rot}\mathbf{H} = j_\Sigma = \left( \varepsilon_0 - \frac{1}{L_{kd}} \frac{1}{\left(\omega^2 - \omega_0^2\right)} \right) \frac{\partial \mathbf{E}}{\partial t} \tag{3.10}$$

$$\operatorname{rot}\mathbf{H} = j_\Sigma = \varepsilon_0 \left( 1 - \frac{1}{\varepsilon_0 L_{kd}} \frac{1}{\left(\omega^2 - \omega_0^2\right)} \right) \frac{\partial \mathbf{E}}{\partial t} \tag{3.11}$$

or

$$\text{rot } \mathbf{H} = j_\Sigma = \varepsilon_0 \left( 1 - \frac{\omega_{pd}^2}{\left( \omega^2 - \omega_0^2 \right)} \right) \frac{\partial \mathbf{E}}{\partial t} \tag{3.12}$$

where

$$\omega_{pd}^2 = \frac{1}{\varepsilon_0 \cdot L_{kd}}. \tag{3.13}$$

Relation (3.13) determines the so-called 'plasma' frequency for the series that are part of the dielectric molecules, if these charges were free.

These results can be obtained without introducing the polarization vector. This is easy to do by calculating current densities. In our consideration, it is assumed that the molecules are in vacuum. In fact, the molecules form a liquid (dielectric solvent). In fact, the density of molecules is very high, but vacuum is still between the two, and bias currents (although they are much smaller than currents associated with the movement of charges) will still be present.

We write the total current density as the sum of current densities bias and conduction current:

$$\text{rot } \mathbf{H} = \mathrm{j}_\Sigma = \varepsilon_0 \frac{\partial \mathbf{E}}{\partial t} + ne\mathbf{v}. \tag{3.14}$$

Using relation (3.13) to find the speed of oscillating charges, we obtain

$$\mathbf{v} = \frac{\partial r_m}{\partial t} = -\frac{e}{m\left( \omega^2 - \omega_0^2 \right)} \frac{\partial \mathbf{E}}{\partial t}. \tag{3.15}$$

Substituting this expression into relation (3.14), we obtain

$$\text{rot } \mathbf{H} = j_\Sigma = \varepsilon_0 \frac{\partial \mathbf{E}}{\partial t} - \frac{n \cdot e^2}{m} \frac{1}{\left( \omega^2 - \omega_0^2 \right)} \frac{\partial \mathbf{E}}{\partial t} \tag{3.16}$$

or

$$\text{rot } \mathbf{H} = j_\Sigma = \varepsilon_0 \frac{\partial \mathbf{E}}{\partial t} - \frac{1}{L_{kd}\left( \omega^2 - \omega_0^2 \right)} \frac{\partial \mathbf{E}}{\partial t}. \tag{3.17}$$

Relations (3.10) and (3.17) completely coincide. Thus, a conditional possibility arises to name the value facing the derivative in the relation (3.12) as the dispersing (frequency-dependent) dielectric constant of the dielectric, which is done currently in all existing literature, which discusses the issues of frequency dispersion in dielectrics. But is such a definition correct? To answer this question, we consider the structure of currents flowing in a dielectric represented by the right-hand side of relations (3.17) and (3.10). The first member of the right-hand side, as was already said, represents the bias current in the vacuum in which the molecules are located. The second term represents the currents associated with the presence in the vacuum of molecules of the dielectric itself. This current is completely identical to the current that occurs in a conventional electrical series oscillatory circuit. Therefore, the equivalent circuit of the unit volume of the dielectric, in which the current distribution can be considered homogeneous, can be represented as a sequential oscillatory circuit with the only difference that the kinetic inductance of the charges should be taken as the circuit inductance. If you take into account the current displacement in vacuum, then in parallel with the circuit it is also necessary to include a capacitance equal to the dielectric constant of the vacuum.

Thus, as in the case of plasma, the coefficient before the derivative in relation (3.12), is not the dielectric constant, but is a composite parameter and now it immediately includes three parameters independent of the frequency. In addition to the dielectric constant of the vacuum of the mind, two more characteristic frequencies enter into it. The frequency $\omega_0$ (let's call it proper) is an individual characteristic of each molecule, it is assumed that it does not depend on the density of filling the space with molecules, and in the space of the molecule located at such a distance that their mutual influence on each other is absent. On the contrary, $\omega pd$ (let's call it the plasma frequency of the dielectric) depends on the packing density of the molecules in the composition of the dielectric fluid. And here arises an important question: what if the composition of a liquid dielectric includes various groups of dissimilar molecules? In this case, each such group will be characterized by its own resonant frequency, while the plasma frequency will be characterized by the density of molecules that make up the dielectric. Now you can imagine which variety of different resonances can be observed in multicomponent systems. This determines the variety of colors that we see around, because at the resonant frequencies there is a maximum reflection or absorption of a signal of a given frequency of electromagnetic waves, which sets off a given colour of the visible us facility. And the purer we see the light, for example in ruby or sapphire, the better the resonance of an atom or molecule whose frequency we observe.

We consider two limiting cases. If $\omega \ll \omega_0$ then from (3.12) we obtain

$$\text{rot } \mathbf{H} = j_\Sigma = \varepsilon_0 \left( 1 - \frac{\omega_{pd}^2}{\omega_0^2} \right) \frac{\partial \mathbf{E}}{\partial t}. \tag{3.18}$$

In this case, the coefficient facing the derivative does not depend on the frequency and represents the static dielectric constant of the dielectric. As you can see, it depends on the natural frequency of the oscillations and on the plasma frequency. This result is clear. The frequency in this case is so small that the inertial properties of the charges do not affect and the magnitude of the polarization vector almost reaches its maximum static values. From the point of view of equivalent electrical circuits, the unit volume of such a dielectric is a capacitance whose value is equal to the coefficient facing the derivative in the relation (3.14).

From the formula (3.18) we can draw another obvious conclusion. The tougher the bonds in the molecule, i.e., the higher the natural frequency, the lower the static dielectric constant. At the same time, the greater their density in space, the higher the static dielectric constant. Immediately she recipe for creating dielectrics with maximum dielectric constant. To achieve this, it is necessary to pack the maximum number of molecules with the most soft bonds between the charges inside the molecule in a given volume of space. Very indicative is the case when $\omega \gg \omega_0$. Then

$$\text{rot } \mathbf{H} = j_\Sigma = \varepsilon_0 \left( 1 - \frac{\omega_{pd}^2}{\omega^2} \right) \frac{\partial \mathbf{E}}{\partial t} \tag{3.19}$$

which corresponds to the transition of the dielectric to the conducting state (plasma), because the obtained relation exactly coincides with the case of the plasma. It was this coincidence that prompted Landau to think that there is no difference between plasma scattering and the behavior of dielectrics at very high frequencies [150]. However, it is not. Indeed, at very high frequencies in dielectrics, due to the inertia of the charges, the amplitude of their vibrations is very small and the polarization vector is also small. At the same time, as in plasma, it is always identically equal to zero regardless of the frequency of oscillations. This consideration showed that such a parameter as the kinetic inductance of charges characterizes any oscillatory processes in material media, be it conductors or dielectrics. It has the same fundamental value as the dielectric and magnetic permeability of the medium. Why has not been seen so far, and why was not it given the proper place? This (again) is due to the fact that physicists are accustomed to think in mathematical categories, without much understanding of the essence of physical processes themselves. From relation (3.3) it is seen that in the case of the equality $\omega = \omega_0$, the oscillation amplitude is infinity. This means there is resonance at this point. An infinite amplitude of oscillations occurs due to the fact that we did not take into account losses in the resonance system, while its quality factor is equal to infinity. In some approximation, we can assume that significantly below the indicated point we are dealing with a dielectric in which the dielectric constant is equal to its static value. Above this point, we are actually dealing with metal, in which the density of current carriers is equal to the density of atoms or molecules in a dielectric. Now, from an electrodynamic point of view, one can consider the question of why the dielectric prism decomposes polychromatic light into monochromatic components. In order for this to take place, it is necessary to have a frequency dependence of the phase velocity (dispersion) of electromagnetic waves in the environment in question. If we add the first Maxwell equation to relation (3.12)

$$\text{rot } \mathbf{E} = -\mu_0 \frac{\partial \mathbf{H}}{\partial t}, \tag{3.20}$$

$$\text{rot } \mathbf{H} = \varepsilon_0 \left( 1 - \frac{\omega_{pd}^2}{\omega^2 - \omega_0^2} \right) \frac{\partial \mathbf{E}}{\partial t}, \tag{3.21}$$

where $\mu_0$ is the magnetic permeability of the vacuum, then from the relations (3.20) and (3.21) it is easy to find the wave equation

$$\nabla^2 \mathbf{E} = \mu_0 \varepsilon_0 \left( 1 - \frac{\omega_{pd}^2}{\omega^2 - \omega_0^2} \right) \frac{\partial^2 \mathbf{E}}{\partial t^2}. \tag{3.22}$$

Given that

$$\mu_0 \varepsilon_0 = \frac{1}{c^2}, \tag{3.23}$$

where $c$ is the speed of light, no one will doubt that during the propagation of electromagnetic waves in a dielectric their frequency dispersion will be observed. But this dispersion will not be connected with the fact that such a material parameter as dielectric constant depends on the frequency. In the formation of such a dispersion, three frequency-independent physical quantities will take part at once, namely: the intrinsic resonance frequency of the molecules, the plasma frequency of the charges, if they are considered free, and the dielectric constant of the vacuum.

In accordance with the concept of a self-consistent field [151], there is such a distribution of the electric field in the system of interacting charged particles that creates a particle distribution that in turn excites this field. A salt solution can be considered as a system of interaction between acting cations, anions, positively and negatively polarized water molecules. Of course, there are also unpolarized water molecules in the solution, but their distribution (to a first approximation) does not affect the distribution of charged particles.

To find a self-consistent field, we still use the Poisson equation

$$\Delta\varphi = -4\pi q \tag{3.24}$$

and the Boltzmann distribution

$$n_k = \tilde{n}_k \exp\left(-\frac{Z_k e\varphi}{kT}\right), \tag{3.25}$$

where $n_k$ is the concentration of particles with charge number $Z_k$ at a point with potential $\varphi$. For electrons, for example, $Z = -1$. But there are no free electrons in the solution. For cations, $Z = m$, and for anions, $Z = -m$, where $m$ is the valency of the metal whose salt is dissolved; $\tilde{n}_k$ in distribution (3.25) is the concentration of particles with a charge number $Z_k$ at a point with zero potential equal to the average concentration of these particles over the entire volume of the solution.

Further, by the index '$i$' we will denote cations, by the index '$a$' – anions, '$p^+$' – polarized and water molecules located around the anions, '$p^-$' – water molecules around the cations.

Average concentrations $\tilde{n}_k$ satisfy the quasineutrality condition

$$\sum_i \tilde{n}_i Z_i + \sum_a \tilde{n}_a Z_a + \sum_{p^+} \tilde{n}_{p^+} Z_{p^+} + \sum_{p^-} \tilde{n}_{p^-} Z_{p^-} = 0, \tag{3.26}$$

which reflects the fact that the solution as a whole (from the outside) is neutral.

The following conditions also apply:

$$\sum_i \tilde{n}_i Z_i = -\sum_a \tilde{n}_a Z_a,$$

$$\sum_a \tilde{n}_a Z_a = \sum_{p^+} \tilde{n}_{p^+} Z_{p^+}, \qquad (3.27)$$

$$\sum_i \tilde{n}_i Z_i = -\sum_{p^-} \tilde{n}_{p^-} Z_{p^-},$$

in which it was taken into account that the charge numbers of the cation and anion are equal in absolute origin, but opposite, and the charge numbers of polarized water molecules are determined by what they are associated around: around anions $Z_{p^+} > 0$, and around cations $Z_{p^-} < 0$.

Thus,

$$\sum_K \tilde{n}_k Z_k = 0, \quad k = i, a, p^+, p^-. \qquad (3.28)$$

The following relationships also occur:

$$Z_a = -Z_i = Z, \; Z_{p^-} = \frac{1}{N_a} Z_{p^-} = \frac{1}{N_i} Z_i, \qquad (3.29)$$

$$\tilde{n}_{p^-} = N_a \tilde{n}_a, \; \tilde{n}_{p^+} = N_i \tilde{n}_i, \; \tilde{n}_a = \tilde{n}_i = n_m,$$

where $N_a$ is the number of water molecules associated around one anion; $N_i$ is the number of water molecules associated around one cation; $Z$ is the valency of a metal whose salt is dissolved; $n_m$ is the concentration of salt molecules in solution, if we assume that they are not dissociated.

The volume charge in this way

$$q = e \sum_k Z_k n_k, \qquad (3.30)$$

where $e$ is the electron charge modulus (in the SI system $1.6 \cdot 10^{-19}$ C). Relation (3.30) is very similar to expression (3.5) with the significant difference that it contains the value of the concentration of particles at the point at which the potential of the self-consistent field is determined.

The Poisson equation is rewritten in the form

$$\Delta\varphi = -4\pi e \sum_k Z_k \tilde{n}_k \exp\left(\frac{Z_k e\varphi}{kT}\right). \qquad (3.31)$$

Using the expansion of the exponential function in a series and taking into account the relations (3.27) and (3.29), we can obtain

$$\sum_k Z_k \tilde{n}_k \exp\left(\frac{Z_k e\varphi}{kT}\right) \approx \frac{Z^2 e n_m \varphi}{kT}\left(2 + \frac{1}{N_i} + \frac{1}{N_a}\right). \qquad (3.32)$$

Even if nothing more than secondary hydration takes place, more than 10 water molecules will be concentrated in the hydration shell. Therefore, we can assume that $2 \gg 1/N_i$ and $2 \gg 1/N_a$. With this in mind, the Poisson equation takes the form

$$\Delta\varphi = -\frac{8\pi Z^2 e}{kT}\cdot\varphi. \qquad (3.33)$$

Solution (3.33) for a spherically symmetric potential distribution around a point charge (cation or anion) has the form

$$\varphi = \frac{C}{r}\exp(-\chi r), \qquad (3.34)$$

where the screening constant is

$$\chi = \left(\frac{8\pi e^2}{kT}Z^2 n_m\right)^{0.5}. \qquad (3.35)$$

Inverse value $l = 1/\chi$ is called the screening length, and we can assume that its value determines the value of the radius of the sphere within which polarized molecules of the solvent line up, and allows us to estimate the value of the radius of the solvated cation. The polarized solvent molecules located within the scope of the electric field of the ion shield it.

Thus, the value of the cluster radius (solvated cation or anion) can be estimated using the relation

$$r_{cl} \approx \left(\frac{8\pi e^2}{kT}Z^2 n_m\right)^{-0.5}. \qquad (3.36)$$

**Table 3.1.** Frequencies for $Y^{3+}$, $Ce^{3+}$, $La^{3+}$ ions at a temperature of 298 K

| Cation | Solvate shell radius $r_{cl}$, μm | Number of $H_2O$ molecules in solvate shell, $\times 10^{14}$ mol | Cooperative rotational movement of $H_2O$ molecules combined into a solvation shell, relative to an axis passing at a distance $R = r_{cl}$ from the center of inertia, $K_r = 1.4$; kHz | Cooperative rotational movement of $H_2O$ molecules combined into a solvation shell relative to the centre of inertia, $K_r = 0.4$; kHz | Frequency of the rotational motion of the cluster as a whole, kHz | The frequency of the transition of the vibrational motion to the rotational one, kHz; $K_r = 0.4$ |
|---|---|---|---|---|---|---|
| $Y^{3+}$ | 12,73 | 2.891 | 0.496 | 1.73 | 0.86 | 0.87 |
| $Ce^{3+}$ | 13.86 | 3.732 | 0.266 | 0.93 | 0.46 | 0.47 |
| $La^{3+}$ | 13.84 | 3.716 | 0.268 | 0.94 | 0.46 | 0.47 |

Table 3.1 shows the frequency values corresponding to various components of the rotational–translational motion of cationic aquacomplexes, and the frequency values corresponding to the transition of vibrational movements into rotational ones. In this case, the cluster radii were determined by the relation (3.36) at a temperature of 298 K and a salt concentration in water of 2 g/l.

It can be seen that, in these approximations, manifestations of the effect of the electroinduced selective drift of cationic aquacomplexes should be expected at electric field frequencies not exceeding units of Hz. Moreover, as follows from relation (3.13), the cluster size is inversely proportional to the square square of the concentration of the salt in the water. The values of frequencies, in turn, are inversely proportional to the value of the moment of inertia of the cluster.

The moment of inertia is proportional to the reduced mass of the cluster, i.e., the number of water molecules in the solvation shell is $g = (r_{cl}/r)^3$, and the quadrature radius of the cluster is $r_{cl}^2$. It turns out that the moment of inertia $I \sim r_{cl}^{-5}$, and the values of the excitation frequencies of the various components $v \sim r_{cl}^{-5}$.

Thus, the frequencies $v \sim n_m^{2.5}$.

It should be expected that with an increase in the salt concentration by a factor of 3–5, the values of the excitation frequencies of the various components of motion, determined in the approximation of existence of self-consistent field in the solution, increase 15–60 times.

The performed experiments prove the possibility of using the previously discovered phenomenon of electroinduced selective drift of solvated ions in salt solutions under the action of the asymmetric electric field for organizing the technological process of enrichment of solutions for the target metal. The experimental results confirm the previously obtained theoretical positions, according to which, at high salt concentrations (of the order of 0.1 g/*l*), the effect induced by an external electric field of selective drift of solvated ions in an aqueous solution is excited at relatively low frequencies. In this case, the frequency interval in which the effect manifests itself at high concentrations is located between the interval characteristic of low salt concentrations [87] and the interval whose boundaries are determined using the radii of solvated ions calculated

by the relation (3.36). It remains to be assumed that the motion of solvated ions, excited by the action of an external asymmetric electric field, is more complex or that the clusters are formed not by one ion and solvent molecules associated with it, but by several ions.

## 3.2. The possibility of forming associates of solvated ions

The value of the cluster radius, obtained in the approximation of the existence of a self-consistent field in the solution volume, raises an obvious question – how does it compare with the characteristic value of the distance between them uniformly distributed in the solution volume?

Simple estimates show that in the case of an aqueous solutionof hydrated yttrium nitrate $Y(NO_3)_3 \cdot 6H_2O$ at a salt concentration of 10 g/l, the average distance between ions evenly distributed in the solution volume is about 100 nm (0.1 μm).

This is significantly less than the estimated values of the radii of the solvation shell (cluster radii) given in Table 3.1. There is no contradiction in this. The fact is that, in the approximation of the existence of a self-consistent field, the value of the screening constant is determined, and the reciprocal of it determines the radius of the ion field, within which the solvent molecules 'respond' to the action of this field. To be more precise, the radius of the cluster should be called the radius of action of a single ion and keep in mind that other ions can also be within this radius.

At high salt concentrations, the radius of action of individual ions overlap, the cluster is formed not by one, but by several ions having a 'generalized' solvation shell.

One can estimate the number of ions within such a generalized solvation shell. For this, it is necessary to determine the electrostatic energy of the associate cluster formed by a given number of solvated ions, assuming that the Coulomb interaction energy of the same charged ions (repulsive forces) is compensated by the interaction energy (attractive forces) of the solvent molecules within the generalized solvation shell.

We will consider the associate as a ball of radius rcl, in the volume of which ni ions, for example, cations, are uniformly distributed. If we calculate the electrical energy of interaction of all ion pairs, considering (for starters) them as points that are approximately uniformly distributed throughout the bulk, then it will be equal to

$$U_1 = \frac{3}{5} = \frac{n_i \cdot (n_i = 1)}{4\pi e_0 r_{cl}} \cdot (z_i \cdot e)^2, \tag{3.37}$$

where $z_i$ is the ion charge multiplicity, $e$ is the electron charge.

We will consider a single ion as a ball of radius $r_i$. The solvent molecule, which is a polar dielectric, is oriented along the centrally symmetric portion of the ion. A molecule is a dipole whose energy in the field of an ion is

$$U = -p_0 \cdot E, \tag{3.38}$$

where $p_0$ is the dipole moment of the solvent molecule: $p_0 = \alpha \cdot \varepsilon_0 \cdot E$, where $\alpha$ is the polarizability of the solvent molecule.

As shown in [6], the molecules of the polar solvent are oriented within a sphere whose radius $r_*$ is determined from the equation

$$\frac{1}{r_*^2} \cdot \ln \frac{C_1}{r^\alpha} \approx \frac{1}{\alpha} \left[ \left( p_0^2 + 2 \cdot \alpha \cdot k \cdot T \right)^{1/2} - p_0 \right], \tag{3.39}$$

where $C_1$ is a constant, the value of which in the SI system is 1 for $Y^{3+}$ and $Ce^{3+}$ ions, the value of the constant $a$ in the SI system for the same ions in the case of an aqueous solution is $-1.951 \times 10^{-10}$. The electric field within this sphere is determined by the relation [6]

$$E_r(r) = \frac{1}{r^2} \cdot \ln \frac{C_1}{r^a}. \tag{3.40}$$

We obtain that the interaction energy (energy of attraction) of one solvent molecule with an ion

$$U_{si}(r) = -\alpha \cdot \varepsilon_0 \cdot E(r)^2 = -\alpha \cdot \varepsilon_0 \cdot \frac{1}{r^4} \left[ \ln \frac{C_1}{r^a} \right]^2. \tag{3.41}$$

The number of solvent molecules in the solvation shell is approximately $N_s \left( \dfrac{r_*}{r_s} \right)^3$.

The energy of attraction of all solvent molecules oriented in the field and 'attached' to the ion:

$$U_s \approx -81 \cdot \frac{\alpha \varepsilon_0}{r_8 \cdot r_s^3} \left[ \ln \frac{3^a \cdot C_1}{r^a} \right]^2 \tag{3.42}$$

In total, the associate includes $n_i$ ions, therefore the energy of attraction of solvent molecules oriented in the field of all ions forming an associate is

$$U_2 \approx n_i \cdot U_s \tag{3.43}$$

The energy of the dipole–dipole interaction (attraction) between two molecules of a polar solvent for the case of a collinear parallel arrangement of dipoles

$$U_{d-d} \approx -\frac{2 \cdot \mu^2}{4\pi\varepsilon_0 d^3}, \tag{3.44}$$

where $\mu$ is the constant dipole moment of the solvent molecule, $d$ is the diameter of the solvent molecule, more precisely, the diameter of the sphere within which an individual solvent molecule is completely located. The energy of the dipole–dipole interaction of all pairs of solvent molecules that are in the associate and do not fall within the spheres of action of the electric fields of individual ions,

$$U_3 \approx -\left( \frac{6}{5} \right)^3 \cdot \frac{n_s \cdot (n_s - 1)}{8\pi\varepsilon_0 \cdot r_{cl}^3} \cdot \mu^2, \tag{3.45}$$

where $n_s \approx \dfrac{r_{cl}^3 - n_i \cdot r_s^3}{r_s^3}$ — number of solvent molecules present in the associate and not field-oriented individual ions. When determining $U_3$, it was taken into account that the average value $\left( \dfrac{1}{r_{ij}} \right)$ for all pairs of points inside a ball of radius $r_{cl}$ is $\dfrac{6}{(5 \cdot r_{cl})}$.

The total electrostatic energy of the associate in a first approximation is defined as the sum

$$U \approx U_1 + U_2 + U_3 \qquad (3.46)$$

It can be assumed that the minimum value of $U$ must correspond to the optimal configuration of the associate formed by $n_i$ ions that are within a sphere of radius $r_{cl}$. The generalized solvation shell of all these ions is formed by solvent molecules, which are also within the sphere of radius $r_{cl}$. The minimum of U is determined by simple differentiation of the general expression for $U$ or for a fixed value of $rcl$, or for a fixed value of $n_i$.

Solution of the $dU/dn_i = 0$ for a given value of $r_{cl}$ allows us to determine the value of $n_i$ corresponding to the minimum of the electrostatic energy of the associate, and solving the equation $dU/dr_{cl} = 0$ for a given $n_i$ allows us to determine the value of $r_{cl}$ corresponding to the minimum of electrostatic energy. The organization of a simple iterative process allows us to determine the optimal combination of a pair of $n_i$ and $r_{cl}$ values, which, in a first approximation, should correspond to the optimal configuration of the associate.

Solution of the $dU/dn_i = 0$ has the form

$$n_i \approx \frac{b + c + d - 2d\left(\dfrac{r_{cl}}{r_s}\right)^3}{2b - 2d\left(\dfrac{r_*}{r_s}\right)^3}, \qquad (3.47)$$

which allows us to determine the optimal number of cations forming associates of radius $r_{cl}$. The constants in the last expression are determined by the formulas

$$b = \frac{3}{5} \cdot \frac{(z_i \cdot e)^2}{4\pi\varepsilon_0 \cdot r_{cl}},$$

$$c = 81 \cdot \frac{\alpha\varepsilon_0}{r_* r_s^3} \cdot \left[\ln \frac{3^a C_1}{r_s^a}\right]^2, \qquad (3.48)$$

$$d = \frac{27}{125} \cdot \left(\frac{r_*}{r_s}\right)^3 \cdot \frac{\mu^2}{4\pi\varepsilon_0 \cdot r_{cl}^3}.$$

For the case of an aqueous solution of cerium salt at concentrations of the order of units g/l, the values of the constants are: $b \approx 10^{-22}$,

$c \approx 5 \cdot 10^{-20}$, $d \approx 3 \cdot 10^{-25}$. For the radius of a cluster (associate) with a diameter of 20 μm, an estimate based on relation (3.47) gives a value of $n_i \approx 10^5$.

The theoretical estimates indicate the possibility of the formation of associate clusters from solvated ions in salt solutions in polar dielectric liquids. It is likely that the action of an external periodic electric field with different amplitudes of intensities in half-periods causes directed motion not of individual solvated ions, but of associate clusters, The significantly larger mass of the associate and, consequently, a larger value of the moment of inertia explains the shift in the range of manifestation of the effect of the electroinduced drift of solvated ions to smaller frequencies at salt concentrations up to 10 g/l, which is in good agreement with experimental results.

When evaluating the frequency values, the coefficient $K_r$ equal to the ratio of the moment of inertia corresponding to the rotation of the solvate shell around the ion to the average projection of the vibrational moment on the axis of rotation. This coefficient characterizes the intensity of the Coriolis interaction [152] (referring to the taxation of Coriolis forces in classical mechanics). It has a value of 0.4 when the solvation shell rotates around the cation (anion) and 1.4 when rotating around an axis passing through relative to the cluster surface. Since the mass of the solvate shell is much greater than its mass, the positions of the centre of gravity (centre of inertia) and the ion practically coincide. The frequency values obtained by the relations (2.9)–(2.11) for $J = 1$ (minimum values) and are given in Tables 2.13 and 3.1, correspond to the case when only one aquacomplex is formed, and allow only an estimate of the frequencies that should be expected in the experiments. As already noted earlier, the size (radius) of the solvation shell is the determining value in assessing the expected value of the frequency at which the manifestation of the effect of the electroinduced drift of solvated ions in the solution is maximum. It, in turn, determines the massus solvate shell and the inertial properties of the solvated ion (cluster). A simple calculation shows that in real weakly conducting liquids with a conductivity of $10^{-10}$–$10^{-12}$ Ω · m, the elementary charge is in a powerful neutral environment, covering $10^{10}$–$10^{12}$ molecules. The mechanism of the ion–molecular interaction remains unclear, especially since weakly conducting liquids are, as a rule, nonpolar liquids that do not have a constant dipole moment. The attached mass of fluid structured around each ion can be estimated by equating the inertia of the ionic structure with the electric force acting on the ion,

$$eE = \gamma V \frac{dv}{dt} \sim \gamma V \omega v_m,$$

where $\gamma$ is the density of the liquid, $v$ is the acoustic velocity of the motion, $v_m$ is the amplitude of the acoustic velocity, $e$ is the electron charge. Under the experimental conditions, $E \sim 10^6$ V/m, $\gamma \sim 10^3$ kg/m$^3$, $f = \omega/(2\pi) \sim 10^5$ Hz, $v_m \sim 10^{-5}$ m/s, cluster volume $V \sim 10^{-18}$ m$^3$, which corresponds to linear dimensions of $10^{-6}$ m. These data indicate that that the introduction of a bulk electric charge into the laboratory fluid is accompanied by the formation of supermolecular formations with physical properties different from the initial medium. If we take into account the electric field of the ions forming the space charge, it turns out that almost all neutral molecules in the liquid are in an inhomogeneous electric field. According to Frenkel, the liquid, in its own way, is closer to crystals than to gases, and retains the so-called 'near order', that is, some ordered arrangement of neighbors is observed near each molecule. Moreover, for liquids with rod-shaped molecules, the ordering increases with increasing molecular length. In the case of ordinary isotropic liquids with rod-like molecules, these groups were discovered experimentally by Stewart [146] using x-ray analysis. Stewart called this phenomenon 'sybotaxia', and groups with an ordered arrangement of molecules called 'sybotactic groups' or 'swarms'.

The intermolecular interaction in homogeneous nonpolar fluids occurs due to the forces of dipole–dipole interactions of molecules with mutually induced dipole moments. In the presence of an ion in a homogeneous liquid in the immediate environment,

This leads to electrical ordering due to the appearance of nonpolar molecules induced by centrally oriented dipole moments. In distant regions, electrical ordering is carried out due to the layer-by-layer induction of dipole moments electrically oriented molecules of layers adjacent to the central ion. The resulting structure is stable, because thermal motion is not able to destroy the electrically oriented structure, consisting of nonpolar molecules. This is argued by the fact that the relaxation time of the induced dipole moments is significant.

It is much shorter than the relaxation time of rotational diffusion of molecules and, therefore, thermal motion cannot affect the arising centrally symmetric structure; therefore, it is stable and quite extended. The spatial orientation of the molecules is also possible due to the anisotropy of the elastic polarizability of the molecules, which is quite significant in the series of normal paraffin hydrocarbons, most of which are organic liquids.dielectrics.

It is believed that the physical nature of supermolecular formations – clusters in nonpolar, weakly conducting media is apparently caused by the ion–dipole interaction in the internal regions of the structure and the layer-by-layer centrally oriented increased dipole–dipole interaction in the external regions. Since the total moments of molecules in nonpolar media are induced, the emerging supermolecular structures are centrally symmetric and are not subject to the disorienting effect of thermal motion. In this case, it is believed that in liquids with constant dipole moments, for example, water, thermal motion destroys supramolecular structures, although this is simply postulated. Which explains such a significant difference in the currently accepted characteristic values of the sizes of solvate shells in aqueous electrolytes and nonpolar weakly conducting media. The theory of aqueous electrolytes is closely related to electrochemistry, in which the 'main actor' is the current of free electrons. Not without reason, we can assume that supermolecular structures – clusters in aqueous electrolytes in the absence of electron current retain their integrity and size. In a polar dielectric, including water, not only layer-by-layer induction of dipole moments by electrically oriented occurs molecules of layers adjacent to the central ion, and layer-by-layer amplification of intrinsic dipole moments. The reason for the destruction of clusters is not chaotic thermal motion (in any case, at low temperatures), but an electron current. Dissolution of salt greatly changes the properties of water. Around each ion, a hydration shell is formed, in which water molecules are ordered differently than in the volume of the solution. In a very dilute solution, we can assume that the structure of the bulk of the solution remains unchanged with individual ion inclusions. If you increase the concentration, then the proportion of volumetric water will decrease and there will come a moment when it will decrease to almost zero - this concentration is called the *boundary of free hydration*. With a further increase in the salt concentration, water molecules will no longer be sufficient even to build complete hydrated shells; finally, the moment will come when all the water in the solution is in direct contact with ions – this is the boundary of complete hydration. It is clear that the structure of such a solution can be very different from the structure of pure water. X-ray diffraction studies showed that it appears to be similar to the structure of the corresponding solid crystalline hydrate. This arrangement of molecules is called a *sybotactic state* (the term was proposed by G. Stewart). At lower concentrations, up to the boundary of free hydration, the so-called sybotactic groups, or sibotactic regions of the solution, are preserved along with bulk water. Therefore, the theory of concentrated solutions cannot be obtained by introducing any corrections into the theory of dilute solutions of Debye–Hückel electrolytes. A complete structural reorganization upon reaching the boundary of complete hydration requires that such a theory be created on fundamentally different postulates.

The set of frequencies corresponding to various components of the rotational–translational and oscillatory motion of the aquacomplex will become even larger if we take into account the nontrivial distribution of the density of solvent molecules in the solvate shell. The above results were obtained by large numbers for a three-dimensional two-particle system, consisting of a hollow spherical shell, in the central cavity of which there is an ion. The density of solvent molecules (solvate groups) in such a shell is assumed to be constant. In reality, the picture is somewhat more complicated. By and large, the solvated ion is a configuration of molecules that make small vibrations around certain equilibrium positions corresponding to the minimum potential energy of their interaction. Its meaning is determined by the expression

$$u = \varepsilon_0 + \sum_{i,k=1}^{r_{\mathrm{sol}}} a_{ik} q_i q_k,$$

where $\varepsilon_0$ is the potential energy of interaction of the molecules when they are all in equilibrium; the second term is a quadratic function of the coordinates determining the deviations of molecules from equilibrium positions. The number $r_{\mathrm{sol}}$ of coordinates in this function is the number of vibrational degrees of freedom of the solvated ion (cluster). The latter can be determined by the number $n$ of molecules in the cluster. It is clear that an $n$-molecular cluster has only $3n$ degrees of freedom. Of these, three correspond to the translational motion of the cluster as a whole, and three correspond to its rotation as a whole. If all the molecules in the cluster are located on one straight line, then there are only two rotational degrees of freedom. Thus, a nonlinear $n$-molecular cluster has only $3n-6$ vibrational degrees of freedom, and a linear one $(3n-5)$. For $n = 1$

the number of the vibrational degrees of freedom is zero since all three degrees of freedom of a single molecule as a whole correspond to its translational motion.

By definition, the total energy $\varepsilon$ of a cluster is the sum of the potential and kinetic energies. The latter is a quadratic function of all pulses, the number of which is equal to the total number of $3n$ degrees of freedom of the cluster. Therefore, the mathematic relation determining the value of the total cluster energy $\varepsilon$, has the form

$$\varepsilon = \varepsilon_0 + f_{11}\,(p,\ q),$$

where $f_{11}(p,\ q)$ is the quadratic function of momenta and coordinates; the total number of variables in this function is $l = 6n-6$ (for a nonlinear cluster) or $l = 6n-5$ (for a linear); for a single molecule $l = 3$, since the coordinates are not included in the expression for energy at all. Substituting this expression for energy in the formula

$$F = -NT \ln \frac{e}{N} \int e^{-\varepsilon(p,q)/T} d\tau,$$

in which integration is performed over the phase space of the cluster, and taking into account that $d\tau = \dfrac{dpdq}{\left(2\pi\hbar\right)^r}$, we obtain

$$F = -NT \ln \frac{e \cdot e^{-\varepsilon_0/T}}{N} \int e^{-f_{11}(p,q)/T} d\tau.$$

In order to determine the temperature dependence of the integral included here, we can substitute $p = p'\sqrt{T}$, $q = q'\sqrt{T}$ for all $l$ variables on which the function $f_{11}(p,\ q)$ depends. Due to the quadratic nature of this function, the equality

$$f_{11}(p,\ q) = Tf_{11}\,(p',\ q'),$$

holds and the temperature $T$ in the index of the integrand will decrease. The transformation of the differentials of these variables in $d\tau$ will give the factor $T^{1/2}$, which can be taken out of the integral sign. Integration over the vibrational coordinates $q$ is performed over the region of their values that corresponds to the vibrations of molecules inside the cluster. Since the integral function decreases rapidly with increasing $q$, then integration can be extended to the entire region from $-\infty$ to $+\infty$, as for all momenta. In this case, the change of variables carried out will not change the limits of integration, and the entire integral will be some constant independent of temperature. Considering also that integration over the coordinates of the centre of inertia of the cluster gives the occupied volume $V$, as a result of free energy we obtain an expression of the form

$$F = -NT \ln \frac{AVe^{-\varepsilon_0/T}T^{1/2}}{N}, \tag{3.49}$$

where $A$ is a constant.

It is possible in a general form to obtain relations determining the thermodynamic characteristics of a solution that is formed by identical clusters.

Differentiating the expression for energy

$$E = F + TS = Nf(T) - NTf'(T)$$

it can be seen that the function $f(T)$ is related to the specific heat $c_v$ by the relation

$$Tf'(T) = c_v.$$

Integrating this relation, we obtain

$$f(T) = -c_v T \ln T - \zeta T + \varepsilon_0,$$

where $\zeta$ and $\varepsilon_0$ are constants. Substituting this expression in

$$F = -NT \ln eV/N + Nf(T),$$

we obtain the following final expression for free energy:

$$F = N\varepsilon_0 - NT \ln eV/N - Nc_v T \ln T - N\zeta T, \tag{3.50}$$

where $\zeta$ is a certain constant characterizing the chemistry of the cluster.

By revealing the logarithm in the expression (3.49), we obtain exactly an expression of the type (3.50) with constant heat capacity equal to

$$c_v = l/2. \tag{3.51}$$

Thus, a purely classical ideal ensemble of solvated ions (clusters) should have constant heat capacity. Expression (3.51) allows us to formulate the following rule: for each variable in the energy $\varepsilon(p, q)$ of the cluster, there is an equal fraction of $1/2$ (in units of $k$, where $k$ is the Boltzmann constant) in the specific heat $c_v$ of the solution or, what is the same, equal share of $T/2$ (in units of $k$) in its energy. This rule is known as the *law of equidistribution*.

Keeping in mind that from translational and rotational degrees of freedom, only the momenta corresponding to them enter the energy $\varepsilon$ $(p, q)$; we can say that each of these degrees of freedom makes a contribution equal to 1/2 to the specific heat. From each vibrational degree of freedom, two variables (coordinate and momentum) enter the energy $\varepsilon$ $(\mathbf{p}, q)$, and its contribution to the heat capacity is 1.

The free energy of the cluster can be represented as the sum of three parts – translational, rotational and vibrational. Due to the large moment of inertia of the cluster as a high molecular weight compound (and, accordingly, the smallness of its rotational quanta) its rotation can always be considered classically. A multimolecular cluster has three rotational degrees of freedom and three generally different main moments of inertia $I_1$, $I_2$, $I_3$; therefore, its kinetic energy of rotation is

$$\varepsilon_{\text{rot}} = \frac{M_\xi^2}{2I_1} + \frac{M_\eta^2}{2I_2} + \frac{M_{\zeta\xi}^2}{2I_3},$$

where $\xi$, $\eta$, $\zeta$ are the coordinates of a rotating system whose axes coincide with the main axes of inertia of the cluster (let us leave aside the special case of a cluster formed by molecules located on one straight line). We substitute the last expression into the statistical integral

$$Z_{\text{rot}} = \int e^{-\frac{\varepsilon_{\text{rot}}}{T}} d\tau_{\text{rot}}, \qquad (3.52)$$

where $d\tau_{\text{rot}} = \dfrac{1}{(2\pi\hbar)^3} dM_\xi dM_\eta dM_\zeta d\varphi_\xi d\varphi_\eta d\varphi_\zeta$,

and integration should be done only according to those cluster orientations that are physically different from each other. If the cluster has any axis of symmetry, then rotations around these axes combine the cluster with itself and are reduced to a permutation of identical molecules. It is clear that the number of physically indistinguishable orientations of the cluster is equal to the number of different rotations it omits around the axes of symmetry (including the identical transformation – by 360° rotation). Denoting this number by $\sigma$, we can to integrate in (3.52) simply in all orientations, at the same time dividing the whole expression by $\sigma$. In the product $d\varphi_\xi d\varphi_\eta d\varphi_\zeta$ (three infinitely small rotation angles), we can consider

$d\varphi_\xi$, $d\varphi_\eta$ as an element $d\sigma_\zeta$ of the solid angle for the directions of the $\zeta$ axis.

Integration over $d\sigma_\zeta$ is performed independently of integration over rotations of $d\varphi_\zeta$ around the $\zeta$ axis itself and gives $4\pi$. After this, integration over $d\varphi_\zeta$ gives another $2\pi$.

Integrating also over $dM_\xi dM_\eta dM_\zeta$ (ranging from $-\infty$ to $+\infty$), as a result we get

$$Z_{\text{rot}} = \frac{8\pi^2}{\sigma(2\pi)}(2\pi T)^{3/2}(I_1 I_2 I_3)^{1/2} = \frac{(2T)^{3/2}(\pi I_1 I_2 I_3)^{1/2}}{\sigma\hbar^3}.$$

$$F = -\frac{3}{2}NT\ln T - NT\ln\frac{(8\pi I_1 I_2 I_3)^{1/2}}{\sigma\hbar^3}$$

Thus, for rotational heat capacity in accordance with (3.51) we have the relation

$$c_{\text{rot}} = \frac{2}{3},$$

and the constant characterizing the chemistry of the cluster is determined by the formula

$$\zeta_{\text{rot}} = \ln\frac{(8\pi I_1 I_2 I_3)^{1/2}}{\sigma\hbar^3},$$

If all the molecules in the cluster are located on one straight line (linear cluster), then it has only two rotational degrees of freedom and one moment of inertia $I$. The rotational heat capacity and chemical constant in this case will be

$$c_{\text{rot}} = 1, \quad \zeta_{\text{rot}} = \ln\frac{2I}{\sigma\hbar^2},$$

where $\sigma = 2$ for a linear multimolecular chain symmetric with respect to the central ion.

## 3.3. Complex vibrations of solvated ions (clusters)

The oscillatory part of the thermodynamic quantities of the solution becomes significant at much higher temperatures than the rotational one, since the intervals of the vibrational structure of the terms are large compared with the intervals of the rotational structure.

Suppose that the temperature is only so high that mostly not too high vibrational levels are excited. Then the oscillations are small (harmonic), and the energy levels are determined by the usual expression $\hbar\omega$ $(v + 1/2)$, where $v$ is the vibrational quantum number.

The number of vibrational degrees of freedom determines the number of so-called *normal* cluster vibrations, each of which corresponds to its own frequency $\omega_\alpha$ (index $\alpha$ numbers normal vibrations). It must be borne in mind that some of the frequencies $\omega_\alpha$ can coincide with each other; in such cases they speak of a multiple frequency.

The calculation of the vibrational partition function $Z_{vib}$ is elementary. Due to the very fast convergence of the series, the summation can be formally extended to $v = \infty$. We agree to count the cluster energy from the lowest ($v = 0$) vibrational level, i.e., we include $\hbar\omega/2$ in the constant $\varepsilon_0$ in the expression

$$\varepsilon_{vK} = \varepsilon_0 + \hbar\omega\left(v + \frac{1}{2}\right) + \frac{\hbar^2}{2l}K(K+1),$$

where $l$ is the moment of inertia relative to the axis passing through the centre of mass of the cluster; $K$ is the quantum number.

In the harmonic approximation, when the oscillations can be considered small, all normal vibrations are independent and vibrational energy is the sum of the energies of each vibration individually. Therefore, the oscillatory statistical sum

$$Z_{vib} = \sum_{n=\alpha} e^{-\hbar\omega v/T} = \frac{1}{1-e^{-\hbar\omega/T}}$$

splits into the product of the statistical sums of individual oscillations,

$$Z_{vib} = \prod_\alpha \frac{1}{1-e^{-\frac{\hbar\omega_\alpha}{T}}},$$

and for free energy $F_{vib}$, the sum of the expressions

$$F_{vib} = NT\ln\left(1-e^{-\frac{\hbar\omega}{T}}\right), \quad F_{vib} = NT \pm \sum_\alpha \ln\left(1-e^{-\frac{\hbar\omega_\alpha}{T}}\right),$$

where $N$ is the number of particles in the system, $T$ is the temperature of the system.

In this sum each frequency is included in the number of times equal to its multiplicity. The same kind of sum is obtained, respectively, for the vibrational parts of other thermodynamic quantities.

Each of the normal vibrations contributes in its classical limiting case ($T \gg \hbar\omega_a$) to the specific heat equal to $c_{vib}^{(a)} = 1$ (in units of $k$, where $k$ is the Boltzmann constant). For a value of $T$ exceeding the largest of $\hbar\omega_a$, we would get $c_{vib} = r_{vib}$, where $r_{vib}$ is the number of vibrational degrees of freedom of the solvated ion (cluster). In fact, this limit is not reached, since the solvated ion (cluster) will decompose into its constituent molecules when temperatures are significantly lower.

The different frequencies $\omega_a$ of the high molecular weight cluster will be distributed over a very wide range of values. As the temperature rises, various normal vibrations will gradually 'turn on' in the heat capacity. This circumstance should lead to the fact that the heat capacity of solutions over fairly wide temperature ranges will remain approximately constant.

So far, we have considered rotation and vibrations as independent cluster movements. In fact, the simultaneous presence of both leads to a kind of interaction between them.

We begin by considering a linear multimolecular chain. Such a circuit can make oscillations of two types – longitudinal with simple frequencies and transverse with double frequencies. Let us analyze the transverse vibrations.

A linear chain of molecules performing transverse vibrations, generally speaking, has some angular momentum. This is obvious from simple mechanical considerations, but can also be shown by quantum mechanical consideration. The latter also makes it possible to determine the possible values of this moment in such an vibrational state of a linear chain of molecules.

Suppose that any one double frequency $\omega_a$ is excited in the molecular chain. The energy level with the vibrational quantum energy $v_a$ is degenerated ($v_a + 1$)-fold. The corresponding wave function is $v_a + 1$ wave functions

$$\psi_{v_{\alpha 1} v_{\alpha 2}} = \text{const} \cdot \exp\left[ -\frac{1}{2} c_\alpha^2 \left( Q_{\alpha 1}^2 + Q_{\alpha 2}^2 \right) \right] H_{v_{\alpha 1}} \left( c_\alpha Q_{\alpha 1} \right) H_{v_{\alpha 2}} \left( c_\alpha Q_{\alpha 2} \right)$$

(where $v_{\alpha 1} + v_{\alpha 2} = v_a$) or any of their independent linear combinations. The common (in $Q_{\alpha 1}$ and $Q_{\alpha 2}$) senior degree of the polynomial by which the exponential factor is multiplied is the same in all these functions and is equal to $v_a$. Obviously, one can always choose linear combinations of the functions $\psi_{v_{\alpha 1} v_{\alpha 2}}$ as the main functions of the form

$$\psi_{v_\alpha l_\alpha} = \text{const} \cdot \exp\left[-\frac{1}{2}c_\alpha^2\left(Q_{\alpha 1}^2 + Q_{\alpha 2}^2\right)\right] \times$$

$$\cdots \quad \times\left[\left(Q_{\alpha 1}+iQ_{\alpha 2}\right)^{\frac{v_\alpha + l_\alpha}{2}}\left(Q_{\alpha 1}-iQ_{\alpha 2}\right)^{\frac{v_\alpha + l_\alpha}{2}} + \cdots\right]. \tag{3.53}$$

In square brackets there is a certain polynomial from which we wrote out only the leading term, $l_\alpha$ is an integer that can take $v_\alpha + 1$ of different values:

$$l_\alpha = v_\alpha, \ v_\alpha - 2, \ v_\alpha - 4, \ \ldots \ , \ -v_\alpha.$$

The normal coordinates $Q_{\alpha 1}$, $Q_{\alpha 2}$ of the transverse vibration are two mutually perpendicular displacements from the axis of the molecular chain. When rotating around this axis through an angle $\varphi$, the leading term of the polynomial (and with it the entire function $\psi_{v_{\alpha 1} v_{\alpha 2}}$) are multiplied by

$$\exp\left\{i\varphi\left(\frac{v_\alpha + l_\alpha}{2}\right) - i\varphi\left(\frac{v_\alpha - l_\alpha}{2}\right)\right\} = \exp(il_\alpha\varphi).$$

.This shows that the function (3.53) corresponds to the state with moment $l_\alpha$ relative to the axis.

Thus, we arrive at a result showing that in a state in which the double frequency $\omega_\alpha$ is excited (with a quantum number $v_\alpha$), the molecular chain has a moment (relative to its axis) running through the values

$$l_\alpha = v\alpha, \ v_\alpha - 2, \ v\alpha - 4, \ldots, \ -v_\alpha.$$

It is regarded as the vibrational moment of a linear molecular chain. If several transverse vibrations are excited simultaneously, then the total vibrational moment is equal to the sum $\Sigma l_\alpha$.

The total angular momentum of the multimolecular system $J$ cannot be less than the moment relative to the axis, i.e., $J$ runs through the values

$$J = |l|, \ |l| + 1, \ldots$$

In other words, states with $J = 0, 1, \ldots, |l - 1|$ do not exist.

In harmonic oscillations, the energy depends only on the numbers $v_\alpha$ and does not depend on $l_\alpha$. The degeneracy of vibrational levels (in terms of $l_\alpha$) is removed in the presence of anharmonicity. Removal, however, is incomplete: the levels remain doubly degenerate, and states differing in the simultaneous change of sign of all $l_\alpha$ and $l$ have the same energy; in the following (after harmonic) approximation the energy shows a quadratic member in moments $l_\alpha$ of the type

$$\sum_{\alpha,\beta} g_{\alpha\beta} l_\alpha l_\beta,$$

Here $g_{\alpha\beta}$ are constants. This remaining twofold degeneracy is removed by an effect similar to $\Lambda$-doubling in diatomic molecules.

Turning to spatially distributed multi-molecular formations, it is necessary first to make the following remark of a purely mechanical nature. For an arbitrary (nonlinear) system of particles, the question arises of how to separate the oscillatory motion from rotation, in other words: what should be understood by 'non-rotating system'. At first glance, one would think that the criterion for the absence of rotation could be the equal to zero angular momentum:

$$\sum m[\mathbf{r}\mathbf{v}] = 0$$

(summation over the particles of the system). However, the expression to the left is not the total time derivative of any coordinate function. Therefore, the described equality cannot be integrated over time so as to be formulated as the equality to zero of some coordinate function. It is this circumstance that makes it possible to reasonably formulate the concept of 'pure oscillations' and 'pure rotation'.

Therefore, in determining the absence of rotation, we must take the condition

$$\sum m[\mathbf{r}_0\mathbf{v}] = 0. \tag{3.54}$$

where $\mathbf{r}_0$ are the radius vectors of the equilibrium positions of the molecules forming the cluster. Denoting by $\mathbf{u}$ the displacements for small oscillations, then $\mathbf{r} = \mathbf{r}_0 + \mathbf{u}$. Integrating (3.42) over time, we obtain

$$\sum m[\mathbf{r}_0\mathbf{u}] = 0. \tag{3.55}$$

Cluster movement we will consider as a set of purely oscillatory motion, under which condition (3.55) is satisfied, and rotation of the cluster as a whole. The translational motion is supposed to be separated from the very beginning by the choice of the coordinate system in which the centre of inertia of the cluster is at rest.

Having written the angular momentum in the form

$$\sum m[\mathbf{rv}] = \sum m[\mathbf{r_0 v}] + \sum m[\mathbf{uv}],$$

we see that in accordance with the definition (3.42) of the absence of rotation, the vibrational moment must be understood as the sum $\sum m[\mathbf{uv}]$. It must be borne in mind, however, that this moment, being only part of the total momentum of the cluster, is not preserved by itself. Therefore, only the average value of the vibrational moment can be attributed to each vibrational state.

The clusters that do not have a single axis of symmetry of more than the second order can be attributed to a many-particle system such as an asymmetric top. For systems of this type, all vibration frequencies are simple (their symmetry groups possess only one-dimensional irreducible representations). Therefore, all vibrational levels are not degenerate. But in any non-degenerate state, the average angular momentum vanishes. Thus, in systems of the asymmetric top type, the average vibrational moment in all states is absent.

If there is one axis of more than the second order among the cluster symmetry elements, it refers to a many-particle system of the type of symmetric top. Such a system has oscillations with simple and double frequencies. The average vibrational moment of oscillations with simple frequencies again vanishes. For double frequencies, there corresponds a nonzero average value of the projection of the moment on the axis.

Expression for the energy of the rotational motion of a cluster of the type of the symmetric top taking into account the vibrational moment has the form

$$\hat{H}_{rot} = \frac{\hbar}{2l_A}\left(\hat{J} - \hat{J}^{(v)}\right)^2 + \frac{\hbar}{2}\left(\frac{1}{l_C} - \frac{1}{l_A}\right)\left(\hat{J}_\zeta - \hat{J}_\zeta^{(v)}\right)^2.$$

The sought energy is the average value of $H_{rot}$. In this expression, the terms containing the squares of the components of the vector $J$ give pure rotational energy. The terms containing the squares of the components of $J^{(v)}$ give constant numbers independent of rotational quantum numbers. They can be omitted. The terms containing the products of the components $J$ and $J^{(0)}$ represent the effect of the interaction of the molecular vibrations with its rotation that is of interest to us. It is called the Coriolis interaction, apparently having a correspondence to the Coriolis forces in classical mechanics.

When averaging these terms, it should be borne in mind that the average values of the transverse ($\xi$, $\eta$) components of the vibrational moment are equal to zero. Therefore, for the average value of the Coriolis interaction energy, we obtain

$$E_{\text{cor}} = -\frac{\hbar^2}{l_C} k k_v,$$

where $k$ (integer) is the projection of the total moment onto the cluster axis, and $kv = \bar{J}_\xi^{(v)}$ is the average projection of the vibrational moment, characterizing this vibrational state; $k_v$, in contrast to $k$, is by no means an integer.

As an example, consider a multi-molecular cluster like a spherical top, that is, a many-particle system of bound particles with the symmetry of any of the cubic groups. Such systems have single, double and triple frequencies. The degeneracy of vibrational levels (as always) is partially removed by anharmonicity. After taking these effects into account, in addition to non-degenerate, only two or three-fold degenerate levels remain. Let's consider these levels split by anharmonicity.

In many-particle systems such as a ball top, the average vibrational moment is absent not only in non-degenerate, but also in doubly degenerate vibrational states. This follows from simple considerations based on symmetry properties. Indeed, the vectors of average moments in two states related to a single degenerate energy level would have to be transformed into each other for all symmetry transformations of the many-particle system. But not one of the cubic symmetry groups admits the existence of two directions that are transformed only into one another. Only sets of at least three directions are transformed into each other.

From the same considerations it follows that in the states corresponding to triply degenerate vibrational levels, the average vibrational moment is nonzero. After averaging over the vibrational state, this moment will be presented as an operator represented by a matrix whose elements correspond to transitions between three mutually degenerate states. In accordance with the number of such states, this operator should be of the form $\xi I$, where $I$ is the moment operator equal to unity (for which $2l + 1 = 3$), and $\zeta$ is a constant characteristic of a given vibrational level. The Hamiltonian of the rotational motion of the cluster

$$\hat{H}_{rot} = \frac{\hbar^2}{2I}\left(\hat{J} - \hat{J}^{(v)}\right)^2$$

after such averaging turns into the operator

$$\hat{H}_{rot} = \frac{\hbar^2}{2I}\hat{J}^2 + \frac{\hbar^2}{2I}\hat{J}^{(v)^2} - \frac{\hbar^2}{2I}\zeta\hat{J}\hat{I}. \qquad (3.56)$$

The eigenvalues of the first term are ordinary rotational energy

$$E = \frac{\hbar^2}{2I}J(J+1),$$

and the second term gives an insignificant constant, independent of the rotational quantum number. The last term in expression (3.56) gives the desired energy of the Coriolis splitting of the vibrational level. Eigenvalues are calculated in the usual way. Energy can have (for a given $J$) three different values (corresponding to the values of the vector $I + J$ equal to $J + 1, J - 1, J$). As a result we find

$$E_{cor}^{(J+1)} = \frac{\hbar^2}{I}\zeta J, \quad E_{cor}^{(J-1)} = -\frac{\hbar^2}{I}\zeta(J+1), \quad E_{cor}^{(J)} - \frac{\hbar^2}{I}\zeta.$$

In many problems of physics, in order to avoid excessive detailing of the systems under consideration, it is assumed that the distribution of one or another physical quantity within a fixed volume of space is uniform (assumption of uniformity). For example, in mechanics, the distribution of matter in the volumes of each of two colliding balls, or the distribution of matter in a body of arbitrary shape, for which it is necessary to determine the position of the center of inertia, is conveniently considered uniform. In electrostatics, the charge distribution in the volume of a charged homogeneous body of arbitrary shape is also conveniently considered uniform. This case can be called paradoxical. The usual charge carrier is an electron whose charge minimum is an elementary charge. The charge of a body of arbitrary shape and arbitrary size can be equal to the charge of an electron. There are no restrictions on the minimum possible value of the body charge. It turns out that the presence of a free electron at any point located in the volume of this body, equally likely at any given time. Nevertheless, the use of the assumption (or assumptions) of uniformity allows us to find a solution to very many practically important problems. There is simply no other way to go over when describing the evolution of material objects from the micro to the macro level, since any attempts to attribute objective physical characteristics to entities of the atomic scale should be abandoned. There is no strict border between the micro- and macro-levels. We ourselves establish it in the formulation and solution of problems of describing spatially separated objects. For example, when we consider the structure of a solution, such a boundary is the size of the solvent molecule. She (the molecule) compares the size and even the shape, as well as the speed of its movement. When we consider the atomic nucleus, the boundary is the size of the nucleons that make up the nucleus. According to Einstein's point of view, the true states of two spatially separated objects are not dependent on each other. This definition of spatial separation cannot be considered simple, since the localization of these systems (objects) according to the Heisenberg uncertainty relation cannot be complete, and the localization regions of two systems in space can overlap. Nevertheless, it is precisely this definition that we use a priori in solving problems of describing the spatial dynamics of bodies of various sizes, starting from electron scattering on atoms and ending with the movement of the planets.

Very often we come across systems that are characterized by two main properties:

1) they are centrally symmetrical;

2) they are a set of identical geometric elements of single elements (units).

An example would be an atomic nucleus in an unexcited state formed by a certain number of equal (in a first approximation) nucleon size. Of course, it is impossible to speak about the size of the nucleon (as such), but it can be said that the wave function of the nucleon has the form of a monochromatic wave within a fixed volume of space, and outside this volume it is equal to zero. This fixed volume is a sphere with a radius equal to the radius of the nucleon. The atomic nucleus can be represented as a system of such spheres tightly adjacent to each other. The core (in turn) can also be considered as a fixed spherical volume of a larger size.

Another example is the solvated metal cation in the solution of salts in a liquid dielectric solvent. It can be represented as a central ion surrounded by solvent molecules tightly adjacent to each other. Assuming that the molecule occupies a spherical volume, it can be said that the solvated cation is a sphere consisting of a certain number of spheres of much smaller size.

In the first case, we can say that the density of matter within each nucleon there is a constant value, and in the second case, that the density of matter within each molecule is a constant value.

In both cases, the system is formed by a set of microspheres (units) identical in size, if we do not take into account that the 'radii' of the neutron and proton are different, and the 'radius' of the central ion is not (in general) equal to the 'radius' of the solvent molecule.

When using the drop model of the nucleus, it is assumed that the density of nuclear matter within the drop is a constant value. When considering the spatial dynamics of a solvated ion, called a cluster, it is also assumed that the density of matter within the cluster is also constant. These are completely conditional assumptions, which allow us to exclude insignificant details from the solution of practically important problems. However, the question of how significant these details are has not been addressed separately. How fair is the assumption that the density of matter in a spherical volume formed by identical spheres of a much smaller volume is constant on average?

Suppose that the system in question is maximally packed, and in its centre is the central microsphere. We call such microspheres units. Thus, the initial and unchanged parameters are the ionite radius $r$ ($d = 2r$) and the density of matter within the ionite $\rho$.

The most compact structure, formed by identical, touching spheres, is either a hexagonal densely packed (HDP) structure or a face-centered cubic (FCC) structure. The fill factor in both cases is 74%. This will be the maximum value, since in both cases we are dealing with maximum packaging.

We single out the cluster in the HDP-structure based on the fact that the central ionite is located in its centre. It will be surrounded by *twelve* nearest neighbours – ionites, whose centres are located at a distance of $2r$ from the centre of the cluster. Together, they form the first layer of ionites equidistant from the centre of the cluster. The second layer of equidistant ionites will be formed by *six* ionites. Moreover, their centres will be at a distance of $2.818r$ from the centre of the cluster. Thus, the cluster will be formed by nineteen ionites.

We distinguish four spherical surfaces, the centre of which coincides with the centre of the cluster, and the radii are equal to: $r$; $1.818r$; $2.818r$ and $3.818r$, respectively. The first of them coincides with the surface of the central ionite, and the last covers the volume occupied by the cluster. Between the 1st and 2nd surfaces there are fragments of ionites of only the first layer, between the 2nd and 3rd – both fragments of ionites of the first layer and fragments of ionites of the second layer, between the 3rd and 4th – fragments ionites of only the second layer.

It is easy to show that the mass of matter concentrated in the range of volumes bounded by spherical surfaces with the indicated radii, will be:

$$m_{un} = \rho \cdot (\pi d^3/6) \approx \rho (0.17\ \pi d^3)^3 \qquad \text{at } 0 \leq r < r,$$
$$m_1 \approx \rho \cdot (0.73\pi d^3) \qquad \text{at } r \leq r < 1.818r,$$
$$m_2 \approx \rho \cdot (1.9\pi d^3) \qquad \text{at } 1.818r \leq r < 2.81r,$$
$$m_3 \approx \rho \cdot (0.37\pi d^3) \qquad \text{at } 2.818r \leq r \leq 3.818r,$$

The average values of the density of matter within the specified limits will be $\rho$; $0.87\rho$; $0.7\rho$ and $0.07\rho$, respectively.

It should be noted that between the first spherical layer (central unit) and the second spherical layer, that is, for $r = r$, the function $\rho$ $(r)$ has a singularity. The first derivative $d\rho$ $(r)/dr$ at this point is discontinuous, and the second derivative does not exist. In the range $0 \leq r < r$ $\rho(r) = \rho$, and starting from $r = r$, the density increases from zero.

If the cluster is distinguished in the FCC structure, then this singularity occurs in the centre of the cluster, i.e., at $r = 0$. Starting from $r = 0$, the density increases from zero. A cluster is a sphere with a central cavity in the centre of which an ion is located.

## 3.4. Radiophysical properties of solutions

Aquacomplexes polarized due to deformation of the solvation shell are located in an external electric field. As a result, there is a separation of aquacomplexes experiencing more and less displacement. The polarization charges of the separated aquacomplexes differ at least in absolute value. Thus, separation of charges and electrostatic forces arise in the volume of the solution. The latter should lead to the excitation of intrinsic electrostatic oscillations into the neutral volume on average (in sufficiently large volumes or for sufficiently large periods of time) of the solution.

As a result of the separation of polarized aquacomplexes, a polarization charge $\rho_{pol}$ arises in solution.

Given the fact that the dielectric susceptibility of the solution is determined by the ratio

$$\chi = \bar{n} \cdot \alpha, \qquad (3.57)$$

where $\alpha$ is the polarizability coefficient of the solvated ion (cluster); $\bar{n}$ is the average number of clusters per unit volume of the solution, and $\bar{n} = 2n_m$, where $n_m$ is the number of dissociated salt molecules per unit volume of the solution, the polarization charge of one cluster $\left( p = \dfrac{\rho_{pol}}{\bar{n}} \right)$ can be determined by the expression

$$p = -\frac{\chi \varepsilon_0}{2 \cdot n_m} \operatorname{div} \mathbf{E} = -\alpha \cdot \varepsilon_0 \cdot \operatorname{div} \mathbf{E}. \tag{3.58}$$

In accordance with the law of conservation of the charge

$$\frac{\partial q}{\partial t} = -\operatorname{div} \mathbf{j}, \tag{3.59}$$

where $\mathbf{j}$ is the current density. Since the current is carried by the polarized aquacomplexes then

$$\mathbf{j} = 2n_m \cdot p \cdot \mathbf{v}, \tag{3.60}$$

where $\mathbf{v}$ is the velocity of aquacomplexes carrying current. The equation of motion of a polarized aquacomplex having mass $m$ is written in the form

$$m \frac{d\mathbf{v}}{dt} = p\mathbf{E}. \tag{3.61}$$

Differentiating equations (3.59) and (3.60) with respect to time and substitution by adding (3.60) to (3.59), we obtain

$$\frac{\partial^2 q}{\partial t^2} = -2n_m p \operatorname{div} \frac{\partial \mathbf{v}}{\partial t}. \tag{3.62}$$

Substituting in (3.62) the expression for $\partial \mathbf{v}/\partial t$ obtained from (3.49), and considering that $\operatorname{div} E = 4\pi q$, we obtain

$$\frac{\partial^2 q}{\partial t^2} = -\frac{8\pi n_m p^2}{m} q. \tag{3.63}$$

The resulting equation describes a simple harmonic oscillation with a circular frequency

$$\omega_0 \sqrt{\frac{8\pi n_m p^2}{m}}. \tag{3.64}$$

If the solution is placed between two flat electrodes, the distance between which is small, then the process can be considered in one-dimensional geometry. Let $E_x$ denote the absolute value of the electric field acting on the solution. Then

$$\omega_0 = \alpha \cdot \varepsilon_0 \cdot \operatorname{div} E \cdot \sqrt{\frac{8\pi h_m}{m}} \qquad (3.65)$$

or

$$\omega_0 = \alpha \cdot \varepsilon_0 \cdot \frac{E_x}{\Delta} \sqrt{\frac{8\pi n_m}{m}},$$

where $\Delta$ is the thickness of the solution layer between the potential and grounded flat electrodes.

Thus, the density of the volume polarization charge in the solution will oscillate with a circular frequency $\omega_0$ ($v_0 = \omega_0/2\pi$) after the action of an external electric field on the solution. In this case, the oscillation frequency is determined not by the amplitude of the field strength, but by the divergence (divergence) of the field strength in the solution volume.

The movements of the water molecule that is part of the solvate shell under the action of an external electric field, as well as the dielectric characteristics of the water structured around the ion, should have features compared to the case of unstructured water. Consequently, the electrophysical properties of the solvation shell formed by water molecules should differ from the properties of ordinary water, since, being part of the shell, water molecules are 'fixed' by the electric field of the ion.

One can describe the polarization properties of a water molecule 'fixed' by an ion.

For this, it is convenient to use a simple model of the molecule – a solid rod, the ends of which carry 'positive' and 'negative' charges, respectively.

### 3.4.1. A water molecule 'attached' to the cation by a hydrogen atom

We assume that the molecule is attached to a positively charged ion (cation) and, to simplify the analysis, is attached to a homogeneous positively charged surface (Fig. 3.1). In the absence of an external electric field, the molecule assumes an equilibrium position, which is the angle $\theta_i$ between the direction of the constant dipole moment of the molecule $\mu_0$ and the direction of the external constant electric field $\mathbf{E}$. When exposed to an external field, the dipole rotates through an angle $\theta$ relative to the equilibrium provisions. An additional dipole moment appears, the projection of which onto the field direction is determined by the relation

$$M = \mu_0 E \sim (\theta_i - \theta). \qquad (3.66)$$

As a result of the deviation of the molecule from the equilibrium state, a quasielastic force $F_{ret}$ arises, which tends to return the molecule back. The molecule will be in equilibrium when the torque of the external electric field is equal to the rotating moment of quasi-elastic force:

$$\mu_0 F \sin \theta = \mu_0 E \sin (\theta i - \theta). \qquad (3.67)$$

The dipole moment arising under the action of an electric field is equal to the difference between the total moment after rotation and the dipole moment before rotation in the absence of an electric field:

$$\mu = \mu_0 \cos (\theta_i - \theta) - \mu_0 \cos \theta_i. \qquad (3.68)$$

If the external electric field is variable, then the molecule will perform forced oscillations described by a linear differential equation of the second order (3.69), in which $\theta$ is taken as a generalized coordinate (see Fig. 3.1):

$$I\frac{d^2\theta}{dt^2} + \beta\frac{d\theta}{dt} + kc = F_{out}l, \qquad (3.69)$$

where $I$ is the moment of inertia relative to the fixing point of the axis of rotation of the molecule. $\beta$ is the coefficient characterizing the internal friction (fluid viscosity), $c$ is the displacement of the dipole under the action of the driving force, and $l$ is the length of the dipole.

The expressions for the driving force (3.70) and displacement (3.71) are determined according to Fig. 3.1 and the sine theorem:

$$F_{dr} = qE \cos\left((90-\theta_i)+\frac{\theta}{2}\right) = qE\left(\sin\theta_i \cos\frac{\theta}{2} - \cos\theta_i \sin\frac{\theta}{2}\right), \qquad (3.70)$$

$$c = 2l \sin\frac{\theta}{2}. \qquad (3.71)$$

Using the Steiner theorem [9], it is possible to calculate the moment of inertia of a water molecule 'fixed' to the cation by an oxygen atom that is part of the molecule. Its value will be $I = 3.0639$

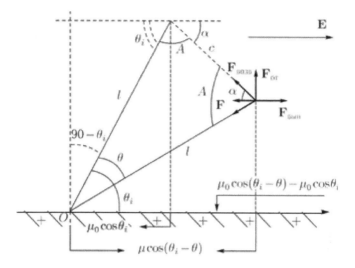

**Fig. 3.1.** The motion scheme of a dipole of a water molecule fixed on a positively charged surface.

$\cdot$ $10^{-47}$ kg $\cdot$ m². The frequency of natural vibrations of the water molecule is $\omega_0 = 1.1851 \cdot 10^{11}$ s⁻¹.

Substituting (3.70) and (3.71) into the equation of forced vibrations (3.69), taking into account the small angle $\theta$, we can obtain the equation of forced vibrations of a water molecule 'fixed' to a solvated cation (positively charged ion) by an electronegative oxygen atom under the action of an external periodic electric field

$$\frac{d^2\theta}{dt^2} + 2b\frac{d\theta}{dt} + \omega_0^2\theta = F_0 e^{i\omega t}\left(\sin\theta_i - \frac{\theta}{2}\cos\theta_i\right). \quad (3.72)$$

The solution to equation (3.72) is

$$\theta = F_0 \sin\theta_i \left[\frac{\omega_0^2 - \omega^2}{\left(\omega_0^2 - \omega^2\right) + 4b^2\omega^2} - i\frac{2b\omega}{\left(\omega_0^2 - \omega^2\right) + 4b^2\omega^2}\right]. \quad (3.73)$$

For the case of dipole oscillations in an electric field, when the angle of the deviation of the dipole from the equilibrium state is small, the expressions for polarizability take the form

$$\alpha' = \frac{\mu_0^2}{2I}\frac{\omega_0^2 - \omega^2}{\left(\omega_0^2 - \omega^2\right) + 4b^2\omega^2}, \quad (3.74)$$

$$\alpha'' = \frac{\mu_0^2}{2I} \frac{b\omega}{\left(\omega_0^2 - \omega^2\right) + 4b^2\omega^2}. \tag{3.75}$$

Dependence graphs of real ($a$) and imaginary ($b$) parts of polarizabilities versus frequency are shown in Fig. 3.2.

The real part of this function has a narrow resonance peak. The position of the maximum of the imaginary part of polarizability is close to $\omega_0$. The peak width and amplitude depend on the attenuation coefficient $b = 1/2\tau$. At low values of $b$, the frequency $\omega_{max}$ tends to the value of the natural frequency of the system $\omega_0$.

The broadening of the spectral line due to an increase in the attenuation coefficient $b$ gradually turns the resonance spectrum into a relaxation one (Fig. 3.3).

Polarization processes in the case of a 'fixed' molecule for elastic electron and ion polarization are similar to the case of a free molecule. For these types of polarization, the 'fixing' method of the water molecule (oxygen atom or hydrogen atoms) to the ion does not matter.

Intermolecular vibrations include librational and translational vibrations of molecules, which correspond to frequencies [154]
$\omega_{04} = 1.2905 \cdot 10^{14}$ s$^{-1}$, $\omega_{05} = 4.0035 \cdot 10^{14}$ s$^{-1}$, $\omega_{06} = 0.3636 \cdot 10^{14}$ s$^{-1}$.

The corresponding polarizabilities are equal to, respectively
$\alpha'_4 = 3.7521 \cdot 10^{-41}$ $\Phi \cdot$m$^2$, $\alpha'_5 = 0.3899 \cdot 10^{-41}$ $\Phi \cdot$m$^2$, $\alpha'_6 = 4.7268 \cdot 10^{-40}$ $\Phi \cdot$m$^2$

The dielectric permittivity for this frequency range has the value

$$\varepsilon' = \varepsilon'_e + \varepsilon'_u + \frac{2n\left(\alpha'_4 + \alpha'_5 + \alpha'_6\right)}{3\varepsilon_0} = 2.81 + 2.23 = 5.04.$$

At the resonant frequency of vibrations of a free water molecule corresponding to the radio frequency region $\omega_{07} = 4.5274 \cdot 10^{12}$ s$^{-1}$, the dielectric constant

$$\Delta\varepsilon = \frac{2n_0\alpha'_7}{3\varepsilon_0} = \frac{2\cdot3.34\cdot10^{28}\cdot0.30486\cdot10^{-37}}{3\cdot8.85\cdot10^{-12}} = 76.$$

The high-frequency dielectric permittivity of a water molecule 'attached' to an oxygen cation corresponds to the value

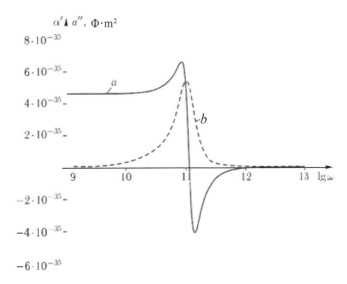

**Fig. 3.2.** Dependence of the real (*a*) and imaginary (*b*) parts of the polarizability on the frequency (Hz).

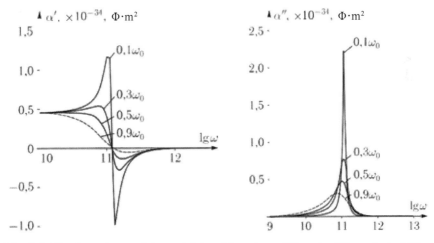

**Fig. 3.3.** The dependence of the real (*a*) and imaginary (*b*) parts of the polarizability on the frequency (Hz) for different attenuation coefficients *b*.

$$\varepsilon' = \varepsilon'_\infty + \Delta\varepsilon = 5.04 + 76 = 81.04.$$

At the natural frequency $\omega_0 = 1.1851 \cdot 10^{11}$ s$^{-1}$, the dielectric permittivity is

$$\Delta \varepsilon = \frac{2n_0 \alpha'_8}{3\varepsilon_0} 11.1937 \cdot 10^4.$$

### 3.4.2. A water molecule 'attached' to the anion by a hydrogen atom

Consider the case of a negatively charged ion (anion). In our simplified model, a water molecule will be fixed on a negatively charged surface by hydrogen atoms (Fig. 3.4). Polarization processes in the case of 'fixing' a water molecule by hydrogen atoms for elastic electronic and elastic ionic polarization are similar to the case of its fixing by an oxygen atom. A significant difference in the values of polarizability and permittivity is observed for elastic dipole polarization.

The moment of inertia of a water molecule fixed on the surface by hydrogen atoms is also calculated by the Steiner theorem and is $I = 11.0385 \cdot 10^{-47}$ kg $\cdot$ m$^2$.

Using the relations (3.74) and (3.75), it is possible to construct the dependence of polarizabilities on the frequency upon attachment by hydrogen atoms and compare it with the dependence for the case of attachment by an oxygen atom (Fig. 3.5).

Consider the behaviour of a molecule at the same frequencies as when fixed by an oxygen atom: $\omega_{01}$, $\omega_{02}$, $\omega_{03}$. The corresponding polarizabilities: $\alpha'_9 = 1.0414 \cdot 10^{-41}$ $\Phi \cdot$m$^2$, $\alpha'_{10} = 0.1082 \cdot 10^{-41}$ $\Phi \cdot$m$^2$, $\alpha'_{11} = 1.3119 \cdot 10^{-41}$ $\Phi \cdot$m$^2$. The dielectric permittivity of water, fixed by hydrogen atoms, in the region of elastic dipole polarization will be

$$\varepsilon' = \varepsilon'_e + \varepsilon'_u + \frac{2n(\alpha'_9 + \alpha'_{10} + \alpha'_{11})}{3\varepsilon_0} = 2.81 + 0.322 = 3.13$$

In the region of radio frequencies, the dielectric permittivity with allowance for $\omega_{07}$ will be $\varepsilon' = \varepsilon'_\infty + \Delta\varepsilon = 3.13 + 21.29 = 24.42$, where $\Delta\varepsilon = 2n_0\alpha'_{12}/3\varepsilon_0 = 21.29$.

At the natural oscillation frequency $\omega_0 = 1.1851 \cdot 10^{11}$ s$^{-1}$, the dielectric permittivity will be

$$\Delta\varepsilon = \frac{2n_0\alpha'_{13}}{3\varepsilon_0} = \frac{2 \cdot 3.34 \cdot 10^{28} \cdot 1.2349 \cdot 10^{-35}}{3 \cdot 8.85 \cdot 10^{-12}} = 0.31 \cdot 10^5 = 3.1 \cdot 10^4.$$

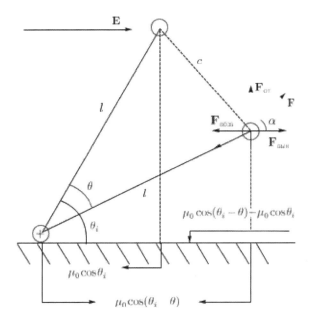

**Fig. 3.4.** The motion of a dipole of a water molecule attached to the surface by hydrogen atoms.

The analysis indicates that the dielectric permittivity of the water structured around the solvated ion, and therefore its electrophysical and radiophysical properties, are significantly different for the cases of cations and anions. In the case of the solvated cation, the water in the solvate shell at the frequency of natural vibrations has an almost 3 times higher dielectric permittivity ($11.2 \times 10^4$) than in the case of the solvated anion ($3.1 \times 10^4$). At the resonant frequency of vibrations of a free molecule of water (4.53 THz), the dielectric constant of water structured around the cation is 76, and for water structured around the anion, it is 21.3. As a result, the high-frequency (in the region of radio frequencies) dielectric permittivity of water structured around the cation is about 81, and around the anion it is about 24.4.

The effect of an electromagnetic wave on the solvated cations and anions in salt solutions in polar dielectrics, all other things being equal, should lead to the excitation of the solvation shells of cations and anions of different intensities, which in principle allows this effect to be used to separate cations and anions when an electromagnetic wave acts on a solution.

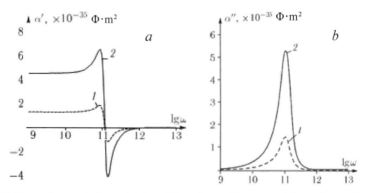

**Fig. 3.5.** Dependence of the real (*a*) and imaginary (*b*) parts of the polarizability on the frequency (Hz): 1 – when fixed by hydrogen atoms, 2 – fixed by an oxygen atom.

## 3.5. Action of an electromagnetic wave on a solution and laser sensing of the structure of a solution

In the absence of a field, gas or liquid molecules are randomly oriented in space. When the laser radiation field is turned on, they are partially oriented along the direction of the external electric field. The axis of the molecule oscillates relative to the direction of the field strength vector. Such a process is called the *alignment* or *orientation* of molecules.

The interaction of a dipole with a strong field of laser radiation leads to the rotation or oscillation of the molecular axis. In the case of a constant dipole moment of the molecule and a constant electric field, this effect is used to orient the molecules along the direction of this field. One important point to make. The behavior of ions with an odd charge in an external field is similar to the behavior of neutral polar molecules such as HCl. Although in the absence of a field ions do not have a dipole moment, a small electric field is enough for an electron to easily move to one of the protons and a constant dipole moment of a molecular hydrogen ion is formed (it is of the order of $eR$, where $e$ is the electron charge, $R$ is the internuclear distance), independent of from the field. The tendency of molecules to line up in the presence of an alternating laser field is well known in nonlinear optics [155]. The alignment of polar molecules is responsible for the so-called orientation Kerr effect, which has been intensively studied in connection with the possibility of orientation of molecules in liquids. Its essence is as follows: the initial optically isotropic medium becomes anisotropic and birefringent under the influence of a constant electric field.

Under the action of the field, the initial indicator of refraction of the medium changes. The refractive indices for linearly polarized light propagating parallel and perpendicular to the direction of the electric field vector become different. The orientational mechanism for establishing optical anisotropy determines the optical properties of the medium. The microscopic nature of the Kerr effect is that a polar molecule interacts with an external electric field and is oriented under the influence of this field so that the energy of the molecule in the field is minimal. This takes place when the dipole moment is oriented in the field. Thermal motion impedes the orientation of molecules. For this reason, the optical properties of the medium depend both on the field strength $E_{el}$ and on the temperature of the medium $T$.

The degree of orientation of the molecules in the electric field depends on the ratio $dE/kT$, where $dE(E_{el})$ is the Stark shift of the energy of the considered electronic state of the molecule in the electric field, which determines the field orientation (see the Stark shift in [156]).

Orientation disappears when this ratio becomes less than unity. On the contrary, thermal motion can be neglected when this ratio is much greater than unity. For example, in accordance with this relation, for a chlorine molecule, thermal motion (at room temperature) can be neglected when the laser radiation intensity is above $5 \cdot 10^{12}$ W/cm². Under such conditions, the task of interaction of laser radiation with molecules is simplified and instead of three-dimensional becomes flat. Indeed, the precession of the molecular axis around the direction of the electric field vector caused by thermal motion is negligible: the molecular axis oscillates only in the plane passing through the initial direction of this axis and the direction of the intensity vector.

The action of an external alternating electric field of the light frequency range on a neutral molecule is similar to the action of a constant electric field, since the orientation time of the molecule is much longer field period due to the large mass of the molecule. The effective value in this case is the average square of the field strength over a period. An exception is the case of a constant dipole moment of the molecule, since then the field perturbation is linear in the intensity of the alternating field and formal averaging over the period makes the perturbation zero. For nonpolar molecules in the light field, the optical Kerr effect takes place. It consists in the fact that atoms and molecules, initially without a constant dipole moment, acquire it under the influence of a field or, as they say, are polarized. The reason for the polarization is electrons, which tend to move in the field in the opposite direction to the electric field, while heavy positively charged nuclei are practically unshifted. In the case of molecules, the anisotropic nature of the induced polarization is due to the initial anisotropy of the structure of the molecule. Since initially the molecules in the medium (in gas or liquid) do not have a fixed orientation, as a result of electron polarization, the nonpolar medium becomes similar to the medium of polar molecules. In the formation of macroscopic anisotropy the arising polarization of molecules is directed along the electric field vector.

The Kerr optical effect is inertialess effect even for picosecond laser pulses (with a duration of the order of $10^{-12}$ s). In almost any molecular medium, the optical Kerr effect arises, which leads to a dependence of the refractive index on the laser field strength and anisotropy of this refractive index relative to the direction of polarization radiation. The task is complicated in the case of femtosecond laser pulses, since then the time to establish the polarizability of the system can be longer than the pulse duration.

In the absence of a laser radiation field, the molecules rotate due to thermal motion. The characteristic rotational quantum numbers can also be either small or large. The classical rotation energy of one diatomic molecule is $kT$. At room temperature, for heavy molecules, chaotic thermal rotation is classical, and for the lightest (for example, a hydrogen molecule, deuterium or their molecular ions) – quantum. Molecular orientation can occur if the molecules of the medium are able to easily rotate.

When a laser radiation field acts on neutral diatomic molecules (for example, HCl), two competing processes can take place: 1) ionization (electron emission) with the formation of a charged molecular ion (in this example, $HCl^+$) and 2) dissociation of a neutral molecule into two neutral atoms (in this example, H and Cl) or two ions (in this example, the proton and ion $Cl^-$). Experiments using linearly polarized light have shown that the dissociation products are detected mainly along the direction of polarization of the laser radiation.

Anisotropy in the distribution of dissociation products can be explained in one of two ways [157]: 1) molecules whose axes are perpendicular to the radiation polarization do not ionize due to the strong angular dependence of the ionization probability, 2) during the dissociation, the molecular axis is oriented along the radiation polarization vector. Angular distributions are determined by polarizability. It is much larger in molecules than in atoms, because in a molecule an electron can move from one nucleus to another under the influence of a field. The polarizability also increases with increasing internuclear distance $R$ in the process of dissociation if charged ions form. Such a strongly polarized system has a large moment of rotation in an external electric field (if it is not already aligned with this field). Although the molecule does not have time to rotate during the ultrashort laser pulse, but it acquires a large angular momentum. This leads to a significant deviation of the trajectories of the products of dissociation and the observed angular distributions of the products of dissociation of light molecules, due to the process of orientation of these molecules. A different situation occurs in the case of heavy molecules. A typical experiment is the effect of linearly polarized laser radiation with an intensity of the order of $10^{15}$ W/cm$^2$ on the iodine molecule. At such a high intensity, for example, the emission of three electrons takes place during a time of the order of atomic times (0.1 fs). The mechanism of emission in such a strong field corresponds to the classical collapse, when the force acting from the side of the laser field exceeds the force that holds the electrons in the molecule. As noted above, the formed molecular ion, as noted above, dissociates into two atomic ions. The Coulomb repulsion between the ions of the same name leads to very fast dissociation, in which the participation of an external laser field is no longer required. This process is called the Coulomb explosion.

It was found in experiments that most molecular ions fly out along the polarization axis of linearly polarized laser radiation. Similar angular distributions take place for the dissociation of other molecular ions formed during the ionization of molecular iodine. The obtained angular distributions are explained by the fact that the probability of multiple ionization depends on the angle between the axis of the molecule and the axis of polarization of the laser radiation. It is greatest when the directions of these axes coincide. The indicated alignment effect is weakened by the fact that the polarizability of molecules and molecular ions is anisotropic. In other words, in addition to the induced dipole moment along the axis of the molecule, there is an (albeit smaller) induced dipole moment in the transverse direction. Actually the longitudinal and transverse polarizabilities differ in one and a half to two times. Dissociation and ionization of molecules in a laser field proceed in a rather complicated way. These processes were considered in [157] as an example of the simplest hydrogen molecule $H_2$.

The laser field cannot increase internuclear distance in a diatomic neutral molecule, consisting of identical atoms, in the form of self-correctness of both directions along the axis of the molecule, along which this distance could increase. Therefore, an external field first pulls out one electron from a neutral $H_2$ molecule. After ionization, a molecular hydrogen ion is formed. The second electron can freely with small internuclear distances move from one nucleus (proton) to another under the influence of the electric field of laser radiation. Since the field oscillates, the electron also easily oscillates between protons with a period equal to the period of radiation. The potential barrier between nuclei is below the energy level of this electron and weakly interferes with such oscillations. At the moment when the electron is near one of the protons, it forms a neutral system with this proton, similar to a hydrogen atom. The laser field does not act on such a neutral system, but on the other hand, it acts on the second proton left without screening, and the increase in the internuclear distance begins. After half a period of radiation, the electron appears at another proton, but the field acting on the proton, now left without screening, changed the sign. Therefore, this field 'pushes' the proton again in the opposite direction to the neutral atom. So, each time there is a repulsion between the charged and neutral systems.

An increase in the internuclear distance occurs to a certain critical value of $R_c$, at which it becomes difficult for the electron to transfer from one proton to another: the effective potential barrier between the protons begins to be interfered with. At this moment, the second electron is ionized, since a further increase in the internuclear distance becomes difficult.

The remaining protons are repelled in a Coulomb way from each other, and their total kinetic energy at infinity is 0. This is confirmed by the available experimental data. The actual critical value of the internuclear distance is three to four times greater than the equilibrium distance. For example, for a molecular ion, the equilibrium internuclear distance is $R_e = 10^{-8}$ cm, and the critical distance $R_c = 3.5 \cdot 10^{-8}$ cm.

So, the neutral molecules can effectively interact with intense laser radiation. This interaction is due to the polarizability of the molecule. The axis of the molecule oscillates relative to the direction of the polarization vector of the laser radiation. The amplitude of the oscillations decreases during dissociation. In the case of light

$H_2$ type molecules, the alignment mechanism is completely different than for heavy type $I_2$ molecules. Effective repulsion between the intradon and the neutral hydrogen atom occurs when the axis of the molecular ion is aligned with the axis of the electric laser radiation vector. This explains the angular distribution of protons with a maximum along the axis of polarization of laser radiation for light molecules, observed experimentally.

The alignment of molecules in the field of laser radiation is stronger for higher intensities of laser radiation and lighter molecules. In the case of heavy molecules, there is no alignment, and during ionization only those molecules are selected whose axes are directed along the polarization of the laser radiation. As in plasma, any separation of charges in a salt solution in a liquid, polar dielectric leads to fluctuations in charge density. On average, over many periods of oscillation, the solution behaves as a quasi-neutral medium. Separation of polarization charges is significant only at time intervals shorter than with a time scale of charge separation $t_0 \sim (\omega_0)^{-1}$. For the spatial scale of charge separation $d$, we can take the distance that the solvated ion travels in time $t_0$ during its thermal motion, i.e., $d \sim \langle v \rangle / \omega_0$, where $\langle v \rangle$ is the average thermal velocity of solvated ions (clusters) having mass m. On a spatial scale, larger than d, quasineutrality of the solution is observed. By definition, the dielectric constant of a continuous medium $\varepsilon$ is the ratio of the strength of the external electric field $E$ to the strength of the weakened field inside this medium $E'$ (dielectric). Moreover, $\varepsilon$ is always greater than 1.

In the solution, $E/E' < 1$. Therefore, for the solution, $\varepsilon < 1$. Moreover, the lower the frequency of the external electric field $\omega$, ($v = \omega/2\pi$), the greater the 'swing' of the oscillations of polarized solvated ions, that is, the amplitude of their displacements. It turns out that with decreasing $\omega$, the dielectric constant of the solution $\varepsilon$ also decreases.

Free electrons in plasma behave in the same way, which is used in the technique of plasma sounding by radio waves. In the case of plasma there is a critical value of the frequency $\omega_k$ of radio waves at which the plasma dielectric constant $\varepsilon = 0$. The value of this frequency coincides with the value of the Langmuir plasma frequency. In our case, the analog of the Langmuir frequency is the frequency of oscillations of the polarization charge in the solution volume $\omega_0$.

The dielectric constant of the plasma is determined by the relation [151]

$$\varepsilon = 1 - \left(\frac{\omega_k}{\omega}\right)^2, \tag{3.76}$$

where $\omega$ is the frequency of external radio emission. If this frequency is $\omega < \omega_k$, then $\varepsilon < 0$.

Maxwell [158] established that the refractive index of an electromagnetic wave in matter $\gamma = \sqrt{\varepsilon}$. For $\varepsilon < 0$, the elec-electromagnetic waves cannot propagate in matter and must be completely reflected from it. Therefore, the plasma is an ideal reflector with respect to waves with a frequency $\omega < \omega_k$. By analogy, we can assume that the salt solution in a liquid, the polar dielectric will be a reflector for waves with a frequency

$$\omega < \alpha \cdot \varepsilon_0 \cdot \text{div } \mathbf{E} \sqrt{\frac{8\pi n_m}{m}}, \tag{3.77}$$

where $\mathbf{E}$ is the electric field in the wave.

From the point of view of physics, the divergence of the electric field vector is an indicator of the extent to which a given point in space is the source or sink of this field:

div $\mathbf{E} > 0$ – the field point is the source;

div $\mathbf{E} < 0$ – the field point is a sink;

div $\mathbf{E} = 0$ – there are no drains and sources, or they compensate each other.

The source of the electric field is the charge, therefore, if the electromagnetic wave is laser radiation, then this charge is the charge induced by laser radiation in the volume of the solution.

In terms of electricity, matter is divided into conductors and dielectrics. The conductors are bodies in which there are free charge carriers, that is, charged particles that can freely move inside this body (for example, electrons in a metal, ions in a liquid or gas). The dielectrics are bodies in which there are no free charge carriers, i.e. there are no charged particles that could move within this dielectric. As previously shown, the salt solution in a liquid polar dielectric at the macroscale is neutral. Each ion is surrounded by solvent molecules that shield the electric field of this ion. The last expression can be rewritten as

$$n_m > \left( \frac{2\pi c}{\lambda \alpha \varepsilon_0 \text{div}\, \mathbf{E}} \right)^2 \cdot \frac{m}{8\pi}, \tag{3.78}$$

where $\lambda$ is the wavelength of electromagnetic radiation, $c$ is the speed of light in vacuum.

We will consider laser radiation as an electromagnetic wave. In this case, the wavelength of the laser radiation is equal to $\lambda$, and the electric field in the laser radiation, as in an electromagnetic wave, is determined by the relation

$$\mathbf{E} = \sqrt{\rho_c \mathbf{P}}, \ \text{V/m} \tag{3.79}$$

where $\rho_c = 120\pi$ is the free-space wave impedance(ohm, $\Omega$), $\mathbf{P}$ is the laser power density (W/m$^2$). The power of laser radiation, including the wavelength of laser radiation, determines the value of $\mathbf{E}$ ($\mathbf{E} \sim \sqrt{\mathbf{P}}/\lambda$) and the divergence of the electric field in the laser radiation div $\mathbf{E}$.

In accordance with (3.78), it should be expected that, for fixed parameters of laser radiation, there is a threshold value of $n_m$ in the salt solution, the excess of which should cause a sharp increase in the reflection coefficient of laser radiation from the salt solution.

The cluster mass $m$ is proportional to the third power of the cluster radius:

$$m \approx 4.163 \cdot 10^3 r^3_{\text{cl}}, \ \text{kg} \tag{3.80}$$

where $r_{\text{cl}}$ is the cluster radius (m); up to a constant $m = a \cdot r^3_{\text{cl}}$

If we conduct experiments in which, for fixed parameters of laser radiation, the reflection coefficient of radiation from a salt solution is measured, then as the concentration increases at a certain value (let's call it critical), we should expect a sharp increase in the reflection coefficient. We denote this value as $C^{\text{crit}}$. The corresponding value is $m_m^{\text{crit}}$. Then the radius of the cluster, formed by the central ion of the salt in the solution can be determined from the ratio

$$r_{\text{cl}} \approx \left( \frac{m_m^{\text{crit}\,1/3}}{\alpha} \right) \left( \frac{\lambda \alpha \varepsilon_0 \text{div}\, \mathbf{E}}{2\pi c} \right). \tag{3.81}$$

A consequence of the Ostrogradsky–Gauss theorem is the equality $\nabla \cdot \mathbf{E} = \dfrac{1}{\varepsilon_0}\rho$, in which the charge density is on the right. Thus, the divergence of the electric field strength is equal to the density of the charge induced by the field in the volume of the solution. The polarization of the solution is equivalent to the appearance of a charge with a density $\rho' = -\text{div } \boldsymbol{\rho} = -\nabla\boldsymbol{\rho}$. This is not obvious. If the polarization vector is constant, then no charge appears in the volume. Now, if the vector changes from point to point, then this is manifested in the fact that a certain fictitious charge $\rho'$ appears in this volume element.

With this assumption in mind, the equation $\nabla{\cdot}\mathbf{E} = \dfrac{1}{\varepsilon_0}\cdot\rho$ can be rewritten in the form $\nabla{\cdot}\mathbf{E} = \dfrac{1}{\varepsilon_0}\cdot(\rho - \nabla{\cdot}\mathbf{P})$ where $\rho$ is the density of real charges, and $-\nabla \cdot P$ is the density of 'coupled' charges, that is, fictitious charges that appear as a result of the polarization of the solution under the action of laser radiation (electromagnetic wave).

We transform the last equation, multiplying everything by $\varepsilon_0$ and carrying $\nabla \cdot P$ to the left. As a result, we obtain the equation $\underbrace{\nabla\left(\varepsilon_0\mathbf{E} + \mathbf{P}\right)}_{\mathbf{D}} = \rho$ where $\rho$ is the density of real charges. Thus, $\nabla\mathbf{D} = \rho$. The vector $\mathbf{D} = \varepsilon_0\mathbf{E} + \mathbf{P}$ is called induction of an electric field generated in the solution volume under the action of laser radiation (electromagnetic wave). Of course, the value of the divergence of the electric field in the laser radiation acting on the solution (divE) is determined not only by the laser radiation parameters, but also by the value of nm. But this issue is a subject of separate consideration.

## Conclusions

A theoretical model of a salt solution in a liquid polar dielectric should take into account the presence of four carriers induced by an external periodic electric field of polarizing charges: solvated cations, solvated anions, positively and negatively polarized solvent molecules.

The action of an electromagnetic wave on a salt solution in a liquid, polar dielectric causes separation of charges and leads to density fluctuations of the polarization charge distributed in the solution volume. Moreover, the prevailing mechanism of separation of metal cations is not the interaction of an electric field with a separate solvated cation, but the collective interaction of an electric field with a polarized charge distributed in the volume.

The radii of the solvation shells of cationic aquacomplexes can be determined in the approximation of the existence of a self-consistent electric field in the solution volume. Theoretical frequency spectrum parameters of complex oscillations of solvated ions theoretically defined as the spectrum of complex vibrations of systems of interacting masses. This allows us to determine the expected values of the frequencies of the electric field at which the manifestation of the effect of the electro-induced selective drift of cationic aquacomplexes is maximized. At concentrations of metal cations of the order of units $g/l$ the sizes of aquacomplexes formed by metals of the third group are micrometers, and the values of the excitation frequencies of the effect are units – tens of hertz.

The theoretical estimates indicate the possibility of the formation of associate clusters from solvated ions in salt solutions in polar dielectric liquids. It is likely that the action of an external periodic electric field with different amplitudes of intensities in half-periods first causes directed motion not of individual solvated ions, but of associate clusters formed by groups of solvated ions. Then a redistribution of the total momentum of the associate between olvated ions having different inertial properties occurs. The significantly larger mass of the associate and, consequently, the larger value of the moment of inertia explains the shift in the range of manifestations of the effect of the electroinduced drift of solvated ions to lower frequencies at salt concentrations up to 10 $g/l$, which agrees well with the experimental results both in the case of cerium and lead nitrates, and in experiments with cerium and nickel chlorides.

# 4

# Prospects for using the effect of selective oriented drift of cationic aquacomplexes in elementary enrichment technology

This chapter discusses electrophysical and electrochemical methods used in elemental and isotopic enrichment substances such as electrodialysis, ionic mobility method. An unconventional approach to solving the problem of processing is presented.

## 4.1. Problem state

As a result of processing ore containing thorium, chemical concentrates are obtained in the form of phosphates, hydroxides, oxalates. They contain from 40 to 70% $ThO_2$, significant quantities of rare earth elements, as well as impurities of uranium, titanium, iron, silicon and others. For example, for thorium concentrates secreted during alkaline treatment of monazite, the following chemical composition is typical, mass. %: $ThO_2$ (50–60); $P_2O_5$ (1.0–2.5); $(REE)_2O_3$ (9.2–15); $SiO_2$ (9.1–30); $Fe_2O_3$ (2–5); $UO_3$ (1.5–2.5); $TiO_2$ (0.1–1.0) - insoluble in the acids of the residue.

The final products of the purification of thorium concentrates – hydroxide and thorium dioxide – should be suitable for use in nuclear engineering. Various methods are used to clean thorium. In this case, thorium concentrates dissolve in sulphuric, nitrogen or hydrochloric acid depending on the method chosen for further cleaning up. If the concentrate contains sparingly soluble compounds of the thorium type phosphates or oxalates, then the concentrate initially treated with a solution of alkali to convert thorium to hydroxide and then dissolving it in acid.

When cleaning thorium compounds, particular attention is paid to the removal of rare earths that have a large cross section of the neutron capture. For cleaning, mainly two groups of methods are used. The first group includes selective precipitation or dissolution methods

- stepwise neutralization;
- selective precipitation of thorium compounds, less soluble, than the corresponding REE salts;
- selective dissolution based on the formation of thorium by soluble complexes with alkali oxalates and carbonates of alkali metals and ammonium.

The second group consists of methods of selective extraction with organic solvents. Currently, the extraction method is of great importance. The methods of the first groups also have not lost their role in industry.

Extraction purification of thorium compounds has great advantages over the methods based on the difference in hydrolytic properties or different solubility, since it allows to carry out deep purification of thorium from most impurities.

Figure 4.1 presents one of the schemes for the extraction purification of thorium [159]. The circuit contains three column type extractors. The supply solution (thorium nitrate solution with a concentration of thorium of approximately 170 g/l, rare earth elements, uranium) is fed into the first extract in which uranium is extracted with a 5% solution of TBF in xylene.

Thorium is extracted from the raffinate of the first extractor in the second extractor with a 40% solution of TBF in xylene. Rare earth elements remaine in the aqueous solution (raffinate). In the third extractor of the circuit, thorium is re-extracted with a solution of 0.02M $HNO_3$. The concentration of thorium in the reextract is approximately 60 g/l. General extraction of thorium according to this scheme from the initial solution reaches 99.8%.

The aqueous raffinate containing REE nitrates is a valuable raw material for the extraction of the required elements.

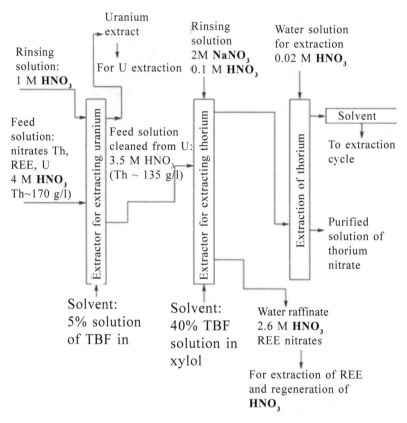

**Fig. 4.1.** Schematic diagram of the extraction purification of a solution of nitrate thorium.

## 4.2. Electrophysical and electrochemical methods in elemental and isotopic enrichment

### 4.2.1. Electrochemical methods of deep cleaning inorganic substances

In solution, macro- and microcomponents can be in a wide variety of chemical forms: in the form of ions and their associates of simple and polymeric neutral molecules, colloidal and fine particles, etc. Therefore, the behaviour of the component management systems in an electric field is not equivalent and depends not only on the physicochemical properties of the solvent and chemical sorts of matter, but also on the nature of electrode processes – reactions of oxidation and reduction, adsorption phenomena, polarization effects, overvoltage, etc. [160].

It should be noted that not all phenomena detected by passing an electric current through a solution have been studied sufficiently for use in inorganic deep purification technology substances. In particular, very interesting in practical terms appears to be the phenomenon of electric transport in liquid metals [161]. It was found that impurities of Na, K, Rb, Cs and Bi in dilute amalgams migrate during electrotransfer to the anode, i.e., they have negative effective charge; impurities Li, Ag, Au, Mg, Zn, Cd and Ga migrate to the cathode. Impurities of Hg (0.4 at.%) and Pb (0.6 at.%) in liquid potassium at a temperature of about 100°C migrate to the anode. It is believed that electric transport is associated with the entrainment of intermetallic compounds of impurities by the electron flux [161]. The possibility of ion separation by the electrogravity method [162, 163], based on the use of convection motion of the concentration-polarization layer at the membrane–solution interface, which creates a concentration gradient along the height of the membrane, has been little studied. This phenomenon was first discovered by W. Pauli [164] during the electrodialysis of inorganic colloids and was called electrostratification.

**Electrodialysis.** Electrodialysis was first proposed as a method of purification of substances by Meigrat and Sabates in 1890, and since then this method has been often used to remove impurities from solutions of non-electric trolites, colloids, and suspensions of sparingly soluble substances [163].

Electrodialysis is a complex physicochemical process occurring in an electrolytic cell separated by semipermeable membranes (diaphragms). The use of membranes causes the emergence of not only new processes, such as dialysis and electroosmosis, but also leads to a change in the number of ion transports in the pores of the membranes along compared with the free volume of the solution, to the appearance of a kind of sorting effect [165]. However, the basis of electrodialysis remains the electrolysis process, the role of which increases with a decrease in the concentration of electrolyte impurities [165, 166]. Using ionite membranes expands the scope of electrodialysis as a method of purifying a substance. There is the possibility of removing trace impurities from electrolytes. In particular, electrolytes containing large ions (e.g. soluble and insoluble polyelectrolytes, salts of large inorganic and organic cations and anions), almost do not pass through ionite membranes, are 'incapable of electrodialysis' [163] and therefore are easily cleaned from trace elements of ordinary electrolytes.

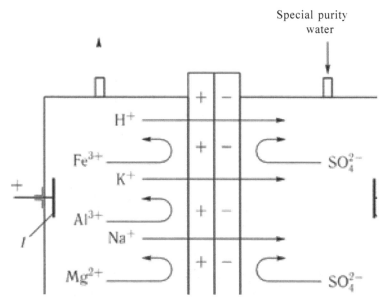

**Fig. 4.2.** Two-chamber electrodialyzer with a bipolar membrane: 1 – anode; 2 – cathode; 3 – anion exchange part of the bipolar membrane; 4 – cation exchange resin part of the bipolar membrane,

The simplest design of an electrodialyzer is an electrolytic cell with a bipolar membrane (Fig. 4.2). Such two-chamber electrodialyzers assembled in a cascade can be used to purify aqueous solutions of salts of divalent and trivalent metals from alkali metal impurities.

Various practical inorganic works of purification of th substances widely use three- (Fig. 4.3) and five-chamber electrodialyzers with narrow and high cameras (to increase membrane surfaces). The electrodializable substance in dissolved form (in the form of a suspension or colloidal solution) is placed into the middle chamber (Fig. 4.3, item 6). In the process of electrodialysis, ions microimpurities are transferred from the middle to the side chambers, which are periodically or continuously washed with very clean water.

For concentrating removed impurities and reducing the consumption of especially pure water it is necessary to use five-chamber electrodialyzers. Additional chambers of these devices are a kind of electric traps of microimpurity ions [166, 167], which prevent back diffusion of the latter into the middle chamber. In that the water is used only for washing additional chambers are washed [167].

For deep cleaning of non-electrolytes rhe electrodialyzers chamber are combined into a cascade called a multi-chamber electrodialyzer

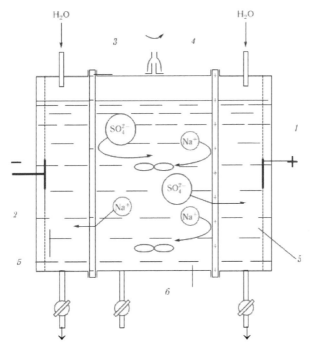

**Fig. 4.3.** Three-chamber electrodialyzer: 1 – anode; 2 – cathode; 3 – cathode membrane; 4 – anode membrane; 5 – side (electrode) chambers; 6 – middle camera..

[163]. In a multi-chamber electrodialyzer, cation exchangers and anion exchange membranes are arranged alternately. If the possibility of penetration of $H^+$ and $OH^-$ ions through the membranes into the middle chamber is not excluded due to non-ideal membranes, this factor in a multi-chamber electrodialyzer affects only those desalination chambers that are in close proximity to the electrode chambers [163].

The electrodes of the electrodialyzers are made only of platinum or highly pure graphite. However, in this case, the possibility of contamination of solutions with electrochemical products of corrosion of electrodes is not excluded.

The efficiency of the electrodialysis process is largely determined by the originally used semipermeable membranes (diaphragms).

In the process of electrodialysis, different sediments ($CaCO_3$, $BaCO_4$, Fe $(OH)_3$, Al $(OH)_3$, $H_2SiO_3$ and others) build up due to the pH gradient formed near the membrane and due to various reactions of the removed ions with the counterions of the membrane, with the presence of colloidal contaminants in the middle chamber or in the initial non-electrolyte, etc. The $Fe^{3+}$, $Al^{3+}$ and $Pb^{2+}$ cations 'poison' most often the cationate membrane. The appearance on the membranes of sediments leads to a decrease in current efficiency and an increase in the resistance of the electrodialyzer. Electrode polarity reversal is used to remove precipitation and flow directions of working solutions and washing water [163].

Therefore, the electrodes must be resistant to the products formed both at the cathode and at the anode. The usual desorption of impurities Fe, Cu, Pb, Cd and others from membranes without reversing the polarity is possible only with prolonged electrolysis with a solution of high acidity.

The negative phenomena observed during electrodialysis include the gradual destruction of the anode membranes due to the release of small amounts (traces) of chlorine, bromine on the anode and oxygen [161].

**The method of ionic mobility**. Inorganic substance purification using the method of ionic mobility (iontophoresis) is based on using minor differences in ion transport numbers of the main component and ions of microimpurities in the electrochemical field. When a sufficiently high potential gradient is combined with a countercurrent of the solvent, a slowdown in the movement of less mobile ions is observed, while more mobile ions go towards the solvent. The less mobile ions seem to be washed off from the more mobile ones [162, 163]. Ion separation efficiency increases with a decrease in diffusion and various convection flows caused by the thermal motion of ions and

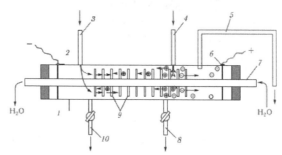

**Fig. 4.4.** The layout of the separation tube for inorganic cleaning substances by the ionic mobility method: 1 – separation tube; 2, 6 – electrodes 3 – tube for supplying a solvent creating a hydrodynamic countercurrent; 4 – tube for supplying the source in the case of removal of cation impurities; 5 – siphon for removal of excess solvent; 7 – pipe water cooling; 8 – conclusion of the solvent, purified from trace impurities; 9 – large porous membranes made of dialysis paper; 10 – output of part of the solvent enriched with microimpurities.

molecules. Therefore, separation tubes (Fig. 4.4), which are the main element of all laboratory facilities using the ion mobility method, are either filled with a fine-grained inert material (silicon dioxide, finely dispersed nozzle made of fluoroplast-4, agar–agar gel and others), or put on cassettes from parallel large-pore membranes that limit the thermal movement of ions and molecules along the solvent stream. The ion mobility method differs from the electrodialysis method only in the absence of in devices and apparatuses (ionophoresisers) of semipermeable membranes. Large-porous membranes used in separation tubes are easily permeable to both anions and cations.

The reaction of two forces – the hydrodynamic pressure of the solvent and the strength of the electric field – leads to the appearance in the separation tube (Fig. 4.4, item 1) of the zones of individual ions in accordance with the values of their carry numbers. The transfer number (relative velocity) of the cation in a solution of a given concentration and temperature is equal to the ratio $n_+ = w_+/(w_+ + w_-)$, where $w_+$ and $w_-$ are the cation and anion velocities, respectively. The ratio of the concentrations ($c_1$ and $c_2$) of ions in two neighbouring zones is equal to the transport numbers $n_1$ and $n_2$ of these ions: $c_1/c_2 = n_1/n_2$. Under normal conditions the boundary of the zones due to the mutual diffusion of ions will always blurry.

The width of the blurred area of two neighbouring zones, $c_1/(c_1 + c_2)$, represents the relative concentration of one of the ions in the zone of another ion at a distance $x$ from the expected interface of the zones. The origin of the $x$ coordinate is taken as the boundary of the zone section ($x = 0$). At this boundary, $c_1 = c_2$ (Fig. 4.5).

Thus, the most effective removal of trace elements by the ion mobility method will be observed only if the substance to be purified is formed by large sedentary ions.

When removing microimpurities similar in physicochemical properties, for the most part, resort to the use of various complexing

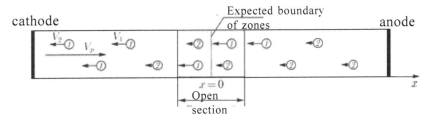

**Fig. 4.5.** Counteraction of hydrodynamic pressure (linear flow velocity $V_p$) of the solvent and electric field strength (linear velocity of ions 1 and 2 under the action of the field $V_1$ and $V_2$, respectively)

reagents that change the mobility of the ions of macro- and microcomponents due to the formation of the latter ionic associates or neutral chelates [164, 165].

The method of ionic mobility (iontophoresis) is a special case of electrophoresis, which is understood as the movement of dispersed electrically charged particles in a liquid medium in an electric field.

Electrophoretic purification is mainly used to remove from non-electrolytes colloidal particles of hydroxides of iron and aluminum; sulfides of arsenic, copper, lead and other metals; oil emulsions (e.g. organic solvent residue after extraction of impurities) and fine mechanical suspensions.

The considered methods can be attributed to contact-current. The electrodes of devices operating on the basis of a particular effect are in direct contact with the working medium and through the boundary of the electrode – medium section flows current. Joule warm to a certain extent affects the efficiency of the process, increasing or decreasing it, and the initiation of the process requires additional costs.

### 4.2.2. HF discharge in elemental and isotopic enrichment

At the end of the 60s, employees of the Sukhumi Physical-Technical Institute discovered the separation of isotopes and gas mixtures in a high-frequency discharge with a running magnetic field. The running a wave, interacting with currents in a gas discharge plasma, compresses the gas in the axial direction. The pressure drop is proportional to the power dissipated in plasma. The sign of the isotope separation effect in the HF discharge is determined by the direction of propagation of the travelling waves. At the end of the chamber where the gas pressure rises, the gas enriched with heavy isotopes. One of the likely mechanism of the isotope separation was calculated by the authors [168] as *thermal diffusion*. It was understood that the radial thermal diffusion effect in neutrals $\varepsilon_R = (\Delta\mu/2\mu) \, R_T \ln (T_{a2}/ T_{a1})$ ($\Delta\mu$ is the difference of the mass number of separated isotopes, $\mu$ is the average atomic weight of the isotope mixture), due to the temperature difference between the wall $T_{a1}$ and discharge axis $T_{a2}$, is converted to longitudinal and y is multiplied due to the internal gas circulation in the discharge (gas circulation occurs due to the radial inhomogeneity of the force $F_z$). The value $R_T$ characterizes the rigidity of the molecules. Calculations showed that under the data conditions of the experiments at an initial pressure

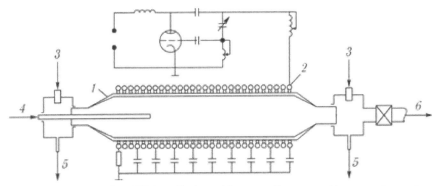

**Fig. 4.6.** Diagram of an RF installation with a traveling magnetic wave: 1 – water-cooled discharge chamber; 2 – delay line; 3 – gauge conversions callers; 4 – filling gas; 5 – gas sampling; 6 – to the pump.

$p > 5.0 \cdot 10^{-2}$ Torr the contribution thermal diffusion in the separation effect is decisive. The authors did not exclude the possibility of the existence of barodiffusion effect at $p < 5.0 \cdot 10^{-2}$ Torr. This pressure range was not in operation investigated in detail.

A facility was built that exceeded the power of the previous one. The results of the experiments on it were published in [169–171]. The installation diagram is shown in Fig. 4.6 [172]. RF discharge was excited in a water-cooled quartz chamber 1, located on the axis of the solenoid of the delay line (Fig. 4.6, pos. 2), consisting of 60 cells. In the delay line, seven-turn coils with a diameter of 0.12 m and ceramic capacitors were used. The length of the discharge chambers $l = 1.1$ m, inner diameter $d = 0.065$ m; connecting nozzles were designed for pumping and sampling gas, as well as sections of the discharge chamber outside the delay line were formed for ballast volumes: $V_L = 1.1 \ l$ and $V_0 = 1.8 \ l$. The volume of the region of the discharge chamber enclosed within the delay line was 2.8 liters.

For experiments with cadmium vapour [171], a thermally insulated discharge chamber with an inner diameter of $d = 0.05$ m was used. The length of the delay line solenoid was $L = 0.85$ m. The delay line was included in the oscillator circuit of the generator (oscillation frequency 80–460 kHz). The phase velocity $V_{ph}$ in the delay line varied in the interval $(0.5–1.5) \cdot 10^5$ m/s. The differential value of pressure $\Delta p = p_L - p_0$ reached $2.5 \cdot 10^{-1}$ Torr ($V_{ph} = 5.5 \cdot 10^4$ m/s, $W = 8$ kW, $f = 80$ kHz). When using a delay line, it is possible to independently change the oscillation frequency and phase velocity of the wave. The discharge is excited in neon, krypton and xenon at initial pressures of $p \leq 2.5$ Torr. The lower limit was determined by the conditions of HF breakdown and corresponded (in Torrs): $5 \cdot 10^{-2}$ (Ne), $4 \cdot 10^{-3}$ (Kr), $1 \cdot 10^{-3}$ (Xe). Discharge was ignited by a preliminary gas ionization using a separate inductor (50 MHz, 100 W). The main results were obtained in experiments with xenon and krypton. A number of experiments were carried out upon excitation of a discharge in a Kr–Xe mixture and also in mixtures of both gases with helium and neon ($p \leq 4 \cdot 10^{-1}$ Torr).

At this facility, a significantly larger separation effect was obtained. The coefficient of enrichment of a mixture of isotopes of xenon ($Xe^{129}$–$Xe^{136}$) exceeded 24%. It is more convenient to characterize the effect by an enrichment coefficient reduced to a unit difference of the isotope masses: $\varepsilon = [(\alpha - 1)/\Delta\mu] \cdot 100\%$. In this case, it is immediately possible to evaluate the separation of any isotopic mixture of a given element. The maximum reduced value is $\varepsilon_{Xe} = 3.5\%$. The increase of the coefficient of enrichment is associated equally with the increase of the power of discharge $W$, and with a decrease in the phase velocity of the wave $V_{ph}$; together in addition, both quantities determine the pressure drop $\Delta p$ and, therefore, ratio $p_L/p_0$ ($\Delta p = W/S \cdot V_{ph}$, where $S$ is the cross-sectional area of the discharge camera). This ratio reached 150 when discharge took place in xenon and 40 when in krypton.

When setting up the experiments, it was assumed that the initial pressure $p \approx \Delta p$ is the boundary pressure: barodiffusion predominates for $p < \Delta p$, and thermal diffusion prevails for $p > \Delta p$.

At low initial pressures ($p \leq 3.0 \cdot 10^{-2}$ Torr), the effect of diluted gases (He, Ne, Kr) on the separation of xenon isotopes is weak. These experiments led to the conclusion that, during a discharge in a gas mixture (low initial pressures), the isotopes of the easily ionized component are more effectively separated [169]. In experiments on the separation of cadmium isotopes [170], xenon is a necessary ballast additive. The isotope enrichment coefficient of cadmium is $\varepsilon_{Cd} = (2.5–3.5)\%$.

The separation coefficients of the mixtures Kr–Xe, Ne–Xe, He–Xe, Ne–Kr were measured. At low initial pressures ($p \leq 3 \cdot 10^{-2}$ Torr), the separation of mixtures is equally effective and is associated with the predominant entrainment of an easily ionized component by a travelling wave. The value of the separation coefficient $\alpha$ is determined by the ratio of partial pressures of the easily ionized component in ballast volumes.

The experiments showed that only isotopes of a component with a lower ionization potential are substantially separated, and the magnitude of the separation effect is determined by the ratio of the partial pressures of this component. Although experiments with a mixture of Ne – Xe at elevated initial pressures were stimulated by the features of thermal diffusion separation of isotopes [173]; it can be noted that an increase in $\varepsilon_{Xe}$ also correlates with an increase in $(p_L/p_0)_{Xe}$.

As for the application of the RF discharge in practice, it seems possible to create a cascade of such plants for the production of some isotopes of cadmium and zinc. But with the achieved level of separation effect, this production will not be profitable, since the energy costs alone are more than $10^5$ kWh/EPP.

## 4.3. An unconventional approach to solving the problem of complex processing of thorium-containing nuclear raw materials and spent nuclear fuel

High temperature helium fluid reactors (HTHFR) are capable of generating heat with a temperature of about 1000°C, which can be used to produce electricity with high efficiency in a direct gas turbine cycle and for supplying high-temperature heat and electricity to hydrogen production processes, technological processes in chemical, oil refining processes, metallurgical and other industries, as well as for desalination of water.

The HTHFR reactors can use both closed and open fuel cycles using uranium, plutonium and thorium. The concept of an open nuclear fuel cycle (ONFC) based on thorium involving uranium and weapon grade plutonium has been tested most extensively on such reactors. The advantages of ONFC based on thorium will increase even more if the cost of producing thorium nuclear material is significantly reduced compared to uranium. This can be achieved either by direct improvement of mining technology and processing of thorium-containing raw materials, or using any features of the genesis and composition of such materials.

If the cost of nuclear fuel can be considered a quantity determined only by the level of technology of its manufacture, then its cost is determined by a number of factors independent of the excellence technological factors. For example, uranium-containing ore mined in the Republic of South Africa is offered on the world market at the lowest price. This fact is not due to the perfection of ore mining technology, but to the fact that it is mined along with gold ore at the same workings. In this regard, some useful thorium-containing mineral fossils secreted monazite. The power and number of deposits of monazite sands can regarded as one of the potential sources of raw materials for large-scale thorium nuclear energy. Monazite (Ce, La, Y, Th)$PO_4$ contains about 12% of thorium dioxide $ThO_2$. Under the action of concentrated acid solutions, for example, nitric or hydrochloric, on the monazite, a mixture of salts of cerium, lanthanum, yttrium and thorium is formed. Modern technological methods allow to separate the thorium concentrate from this mixture in the form of this or other chemical compound. The remaining pulp represents a very valuable raw material to obtain either in the form of compounds, or in the pure form of rare earth metals. Thus, it is possible to implement a technological option in which the thorium concentrate is an associated material in the preparation of cerium, lanthanum and yttrium concentrates.

Technology will be based on the phenomenon of induced selective drift of cationic aquacomplexes in salt solutions under the action of asymmetric electric fields whose frequency does not exceed tens of kilohertz.

The separation element diagram of the technological cell is shown in Fig. 4.8.

The input of the element (pipe for supplying solution 1, Fig. 4.7), receives the initial mixture of 2 components (an aqueous solution of a mixture salts $Ce(NO_3)_3$ and $Y(NO_3)_3$ with concentrations of 3.5 and

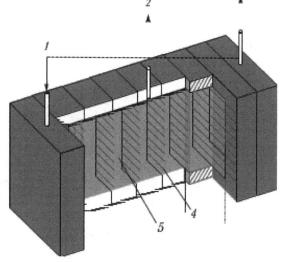

**Fig. 4.7.** Technological cell: 1 – pipe for supplying the solution; 2 – nozzles for sampling the solution; 3 – pipe for the selection and organization of forced circulation of the solution; 4 – potential, isolated from solution of the grid; 5 – solution.

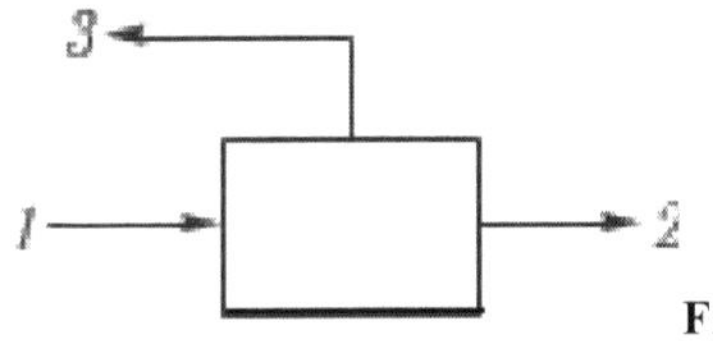

**Fig. 4.8.** Scheme of the separation element.

3 g/l, respectively – the feed stream $L$ with concentrations $C_i$ and $C_j$, respectively. Two streams emerge from the element: 2 – selection of $L'$ (aqueous solution, enriched with $Ce^{3+}$, branch side 2, Fig. 4.7) and 3 – dump $L'$ (water solution enriched from $Ce^{3+}$, branch side side 3, Fig. 4.7).

The concentrations of the components in the selection are equal: $Ce^{3+} - C_i'$; $Y^{3+} - C_j'$ and in the dump $C_i''$ and $C_j''$, respectively. In this case, the selection is enriched with $Ce^{3+}$ cations and depleted in $Y^{3+}$ cations, and the dump, on the contrary, is depleted in $Ce^{3+}$ cations and enriched in $Y^{3+}$ cations.

Table 4.1 shows the experimentally determined values of quantities in the accepted notation.

Thus, after 4 hours of exposure to the field with intensity $E^+ = 14.3$ V/cm, the asymmetry coefficient $A^-/A^+ = 0.66$, with a frequency of 1.6 kHz per aqueous solution of a mixture of salts of $Ce(NO_3)^3$ and $Y(NO_3)_3$ the values of the concentrations in the absence of circulation of the solution through the separation element make up: $C_i' = 1.00249$; $C_j' = 0.99751$; $C_i'' = 1.00000$; $C_j'' = 1.00179$. The relative concentrations of the cations of cerium and yttrium at this comprises: $R_{ij} = 1.00000$ (stock solution); $R_{ij}' = C_i'/C_j' = 1.00499$ (selection) and $R''ij = C_j''/C_i'' = 0.99821$ (dumping).

The separation factors in an element without solution circulation make up:

- $\alpha_{ij} = R_{ij}'/R_{ij} = 1.00499$;
- in the dumping $\beta_{ij} = R_{ij}/R_{ij}'' = 1.00179$;
- full $q_{ij} = \alpha_{ij} \cdot \beta_{ij} = 1.00679$.

Thus, it is possible to implement a technological option in which the thorium concentrate is a by-product with obtaining concentrates of cerium, lanthanum and yttrium.

**Table 4.1.** Concentrations of components in accepted designations

| Concentration, arb. units | Feed flow | Sampling | Dumping |
|---|---|---|---|
| $C_i + C_j$ | 0.99284 | 1.00000 | 1.00179 |
| | 0.99254 | 1.00249 | 1.00000 |

194

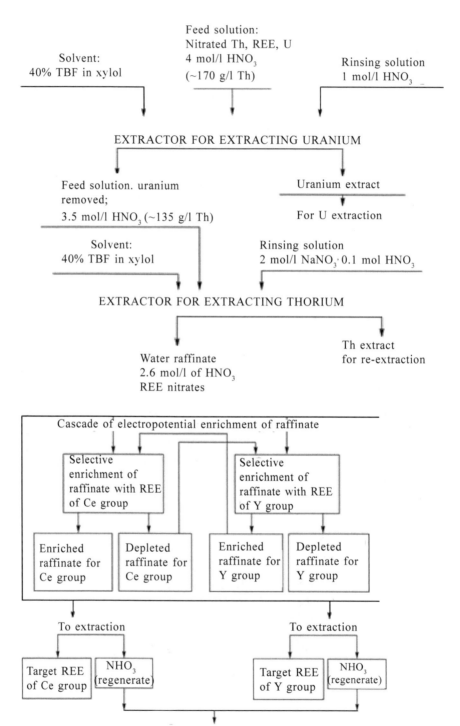

**Fig. 4.9.** Modified system of extraction purification of solutions of Th nitrites.

Figure 4.9 shows a modified scheme of extraction purification of a solution of thorium nitrate, allowing to extract from the aqueous raffinate of the target REE. A well-proven process flow diagram adds a new link to extract industrially significant amounts of the target REE (e.g., yttrium or cerium).

In the case of complex hydrochloric acid technologies for the rare metals, the schematic diagram of the extraction of REE from aqueous raffinate containing hydrochloric acid and REE chlorides will not change. Of course, the optimal combination of electric field parameters of the mixture of chlorides present in an aqueous solution will be different.

In this case, the orders of magnitude of the electric field strength and its frequencies will not change. Separation cell design, the circuitry and power of the high-frequency asymmetric electric field generation system will remain unchanged. Thus, the developed technology is universal in relation to the scheme of extraction purification of thorium and to the technology of rare-metal raw materials.

Technology is not literally electrochemical in direct meaning of its term, since mass transfer occurs in solutions electrically insulated from electrodes onto which the voltage source is 'loaded'.

.These features and the fact that the amplitude values of the electric field at which the phenomenon occurs are units of volts per centimeter, allow one to attribute the technology to energy-saving.

Spent nuclear fuel (SNF) contains a significant amount of platinoids whose atomic nuclei are fission fragments of fuel cores. Their content in SNF depends on the type of reactor, type and depth of burnout, SNF holding time and a number of other parameters. For thermal neutron reactors with $UO_2$ fuel and the depth of burnout of 33 GW $\cdot$ day/t after 10 years of ageing on average per ton the fuel accumulates 2.1 kg of ruthenium, 0.4 kg of rhodium and 1.2 kg of palladium. In the fuel of fast neutron reactors, the accumulation of these metals increases by an order of magnitude. Thus, SNF for the content of platinum metals should be considered, and is currently considered by experts as an alternative source of these metals.

Without changing the time-tested technologies for SNF processing, one can supplement them with one or more links in which dividing cascades will operate, allowing extraction of valuable metals from aqueous solutions of salts. The results of studies show the possibility and feasibility of using the effect of electroinduced selective drift

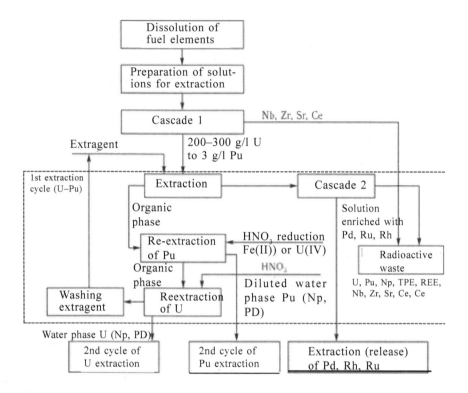

**Fig.. 4.10.** Improved scheme of the first extraction cycle of the Purex process.

in the traditional Purex process. Separation cascades based on the effect of electroinduced selective drift, are supposed to be introduced into the general technological scheme of the Purex process in two stages (Fig. 4.10).

The first separation cascade is supposed to be introduced before the first extraction cycle to extract nitrate from the composition of the nitric acid solution of the most active radionuclides. These include fission products like Zr, Nb, Sr, Ce and other REEs. Removing the listed fission products, which are hard $\gamma$-emitters, not only reduces the radiation load on the extractant, increasing thereby its capacity, but also reduces its destruction due to the action of $\gamma$ radiation. This measure will significantly reduce radiation effects on chemicals and solvents in the process of reprocessing the irradiated nuclear fuel therefore reducing the amount of medium-level waste and reducing the set of elements from which basic products should be cleaned. The use of the second separation cascade in the technological scheme of Purex-process is supposed to be extracted from nitric acid solutions (water-tail solutions) after the first extraction cycle of the platinum metals.

## 4.4. Effects in biotechnology and medicine

Water makes up the base of the blood stream, cytoplasm and intercellular fluid. In all of these fluids, in varying concentrations, inorganic substances, proteins, sugars, and cell formations are present that are hydrated and have the appropriate hydration shell. Blood can be considered as a heterogeneous multicomponent system containing red blood cells, white blood cells, platelets, which are suspended in a colloidal solution of the electrolytes, proteins and lipids. Almost all blood components are hydrated, and their rotational mobility plays an important role in the bioenergy of the body. Rotational mobility as was shown above, is controlled by an external asymmetric electric field of low tension, which allows the latter to be used for selective exposure to certain blood components, cytoplasm or intercellular fluid, enhancing or inhibiting the mobility of certain components. Selective drift effect of the solvated ions can also be used in the technology of purification of blood from harmful compounds, each of which is characterized by a specific value of the rotational transition frequency into the translational component of motion.

The defining feature of all biologically active molecules is their chirality. Molecules can be considered as geometric bodies. If there are no symmetry of the elements of the Sn groups in the molecule, then such a molecule is chiral. Similar molecule and its mirror images are pairs of isomers that are not compatible with each other. Such isomers are called mirror antipodes, or enantiomers. (enantiomers are synonymous with chiral molecules). For example, a methane molecule is achiral (see Fig. 4.11), so as its mirror reflection is fully compatible with the original image during movement, and the bromofluorochloromethane molecule is chiral, so both she and

**Fig. 4.11.** Molecule of methane is achiral (6 planes of symmetry σ, 3 axes symmetry S4).

**Fig. 4.12.** .Chromium bromofluorochloromethane molecules – pair of enantiomers.

her mirror image are pairs of isomers (see Fig. 4.12) that do not fit together. In doing so, *the physical properties of the enantiomers are identical.* A mixture of enantiomers is called a racemate. So, during the synthesis of bromofluorochlomethane, molecules of two geometric shapes can form. As a result the synthesized bromofluorochloromethane will be a mixture of two enantiomers, i.e. a racemate.

Chirality is crucial in the synthesis of complex compounds with pharmacological properties. Since natural sources can no longer satisfy the existing need for biologically active compounds, the method of cleavage of racemates, i.e. separation, is used to obtain enantiomers included in the racemic mixture. There are 3 ways of cleavage of racemates: mechanical, biochemical and chemical. The chemical method is currently the main method of separation of racemates. Its essence is the translation of both enantiomers into diastereomers, the physical properties of which already differ, with their subsequent separation.

If we consider the solution of the racemate in the approximation of the existence in its volume of a self-consistent electric field, then in it there are 3 carriers of a polarization charge induced by an external nelectric field – solvent molecules and solvated enantiomer molecules. The action of an external asymmetric electric field on the solution of the racemate will lead to the excitation of a rotational–translational motion of the solvated enantiomer molecules, and differences in the drift parameters of each of the enantiomers will result to their separation in space and will provide the possibility of their selective extraction from solution. Thus, the effect of selective drift can be used in the synthesis of biologically active substances or for a controlled change in the properties of their solutions.

## Conclusions

The electrophysical and electrochemical methods are considered: interchangeable in elemental and isotopic enrichment of substances, such as electrodialysis, ion mobility method, as well as the use of RF electric field. Since all of them are based on the action of an electric current, the technologies for producing OF and monoisotopic substances based on them are energy intensive. In this regard, the development of energy-saving technologies and their natural-scientific foundations is conditional. Development of the natural scientific foundations of technology of elemental enrichment of aqueous solutions of a mixture of salts under the action of an asymmetric high-frequency electric field can become an integral part of a program for developing a complex of technologies for promising types of nuclear fuel, in technologies for purifying blood from harmful compounds, for synthesizing biologically active substances and for controlled changes in the properties of their solutions.

The discovery of the phenomenon of electroinduced selective drift of cationic aquacomplexes in aqueous salt solutions raised many questions, one of which was the question of the structure of the salt solution in a polar dielectric liquid and on the structure of clusters formed by ions and solvent molecules associated around them. Traditional models and approximations, in which many interesting results were obtained, did not give the opportunity to quantify the estimate correlating with experimentally observed effects. This triggered a search for new models and approximations that were laid down in the monograph and partly explain the physical nature of the detected phenomenon. Presumably, the search will continue and the developed models and accepted approximations are subject to further improvement. Nevertheless, the possibility of using the discovered patterns of interaction of external unsteady electric fields with salt solutions to create and development of a new, electrodeless, energy-saving technology enrichment of solutions with target elements has already been indicated.

The development of the scientific foundations of this technology has shown that the inertial properties of the supramolecular structural units of clusters are determined by their size. The sizes of these clusters range from tens of angstroms up to several microns. At the same time, liquid media are the basis for the functioning of both biological and technical systems. Thus, the aqueous salt solution is formed by nanostructures, and biological and many technological systems are functioning with the participation of nanoparticles. (e.g. conventional circulatory system or many of the chemical processes). Random or continuous action of aperiodic electric fields, magnetic fields and electromagnetic waves on systems containing salt solutions, causes either positive or negative effects. Very often we encounter them, but do not pay due attention finding out their reasons.

The technology in modern conditions of Russia can be considered new if:

• there are no other technologies for the production of one or another product

• or it differs from its counterparts by an advantage (a number of advantages) in some parameter (a number of parameters), provided that in none of the other parameters it is inferior to analogues;

• or it is based on a principle (physical phenomenon, effect, sequence or combination of actions) that has not previously been used in engineering and technology.

The book shows the current state of research found by the authors of an effect whose applications will help ensure the transition from the stereotype 'the more expensive, the safer' to the norm 'the safer the cheaper'.

The above allows us to conclude that the electrophysics of structured solutions of salts in polar dielectric liquids should be an integral part of research conducted in the field of nanotechnology, biology, medicine and, of course, in the field of technology of highly pure substances. The authors hope that the problems discussed in the book will contribute to the development of nonequilibrium molecular physics as a science of physical and chemical processes, as well as related disciplines related to structural and phase transformations in biomedical solutions, in particular, in the field of recording and transmitting information in liquid media.

# References

1. Theoretical physics: Textbook: In 10 volumes / L. D. Landau, E. M. Lifshits. - 3rd ed., Rev. - M.: Nauka, 1992. T. 8: Electrodynamics of continuous media. - 664 p.
2. Fundamentals of the theory of electricity: a training manual / I. E. Tamm. - 11th ed., Rev. and add. - M .: Fizmatlit. 2003 .-- 615 p.
3. Kazaryan M. A., Lomov I. V., Shamanin I. V. Electrophysics of structured solutions of salts in liquid polar dielectrics. - M .: Fizmatlit, 2011 .-- 192 p.
4. Feynman R. F. Feynman lectures on physics: Per. from English / R.F. Feynman, R. Leighton, M. Sands. - 3rd ed., Revised. - M.: Mir, 1976-1978.
5. Parsell E. Berkeley Physics Course: Trans. from English: In 5 vol. 3rd ed., rev. T. 2: Electricity and magnetism. - M.: Nauka, 1983.
6. Kazaryan M. A., Shamanin I. V., Lomov I. V. et al. Electrical and magnetically induced transport of solvated ions in an isolated salt solution in a polar dielectric // Theoretical Foundations of Chemical Technology, 2010. T 44, No. 1, pp. 1–9.
7. Kamke E. Handbook of ordinary differential equations: Per. with him. / E. Kamke. - 5th ed., Stereotype. - M.: Nauka, 1976.
8. Shamanin I. V., Kazaryan M. A., Gofman V. N. et al. Separation of solvated calcium and magnesium cations by the action of an external periodic electric field on a moving solution // Bull. Lebedev Phys. Inst. (2017) 44: 137. P. 23–32. DOI 10.3103 / S1068335617050049
9. Madelung E. Mathematical apparatus of physics: a reference guide. - M .: Nauka, 1968 .-- 620 p.
10. Gusev A. L., Kazaryan M. A., Lomov I. V., Trutnev Yu. A., Shamanin I. V. Structuring of solutions in polar dielectric liquids and separation of solvated ions under the action of an external asymmetric electric field / / Alternative energy and ecology, No. 06/2, 2013. S. 10–22.
11. Gusev A. L., Kazaryan M. A., Lomov I. V., Trutnev Yu. A., Shamanin I. V. Effect of an external asymmetric electric field on salt solutions in dielectric liquids: physics of the process and applications // Alternative Energy and Ecology, No. 05/2 (126), 2013. P. 35–45.
12. Landau L. D., Lifshits E. M. Theoretical Physics: T. 1. Mechanics. - M .: Nauka, 1988 .-- 216 p.
13. The method of large particles in gas dynamics; Computational experiment / O. M. Belotserkovsky, Yu. M. Davydov. - M .: Nauka, 1982.- 391 p.
14. Paskonov V. M., Polezhaev V. I., Chudov L. A. Numerical modeling of heat and mass transfer processes. - M .: Nauka, 1984. - 288 p.
15. Kazaryan M. A., Shamanin I. V., Dolgopolov S. Yu. Et al. Electro-and magnetically induced transport of solvated ions in an isolated salt solution in a polar dielectric // Theoretical Foundations of Chemical Technology. 2010.Vol. 44, No. 1. P. 1–9.
16. The basic laws of atomic and nuclear physics: Textbook / E. A. Nersesov. - M .: Vysshaya shkola, 1988 .-- 287 p.
17. Erdei-Gruz T. Fundamentals of the structure of matter / Ed. G. B. Zhdanova. - M.:

Mir, 1976 .-- 488 p.

18. Stishkov Yu. K., Steblenko A. V. Violation of the homogeneity of weakly conducting liquids in strong electric fields // Journal of Technical Physics. 1997. Vol. 67, No. 10. P. 105–111.

19. Stishkov Yu. K., Ostapenko A. A. Electrohydrodynamic flows in liquid dielectrics. - L .: Publishing House of Leningrad University, 1989 .-- 174 p.

20. Karapetyants M. Kh., Drakin S. I. The structure of matter. - M.: Vysshaya shkola, 1970. - 312 p.

21. Lukovsky I. A., Chernova M. A. Non-linear modal theory of droplet vibrations // Acoustic Bulletin. 2011. Vol. 14, No. 3. P. 23–45.

22. Landau L. D. A short course in theoretical physics: a training manual: in 3 books. / L. D. Landau, E. M. Lifshits. - M.: Nauka, 1972.

23. Ghazaryan MA, Shamanin IV, Melnik N. N. et al. Structure and radiophysical properties of salt solutions in liquid polar dielectrics // Chemical Physics. 2009.Vol. 28, No. 2. P. 20–26.

24. Shamanin I. V., Kazaryan M. A. Structure of salts solution in polar dielectric liquids and electrically-induced separation of solvated ions // Proceedings of the 13th Workshop on Separation Phenomena in Liquids and Gases (SPLG 2015). Bariloche, Argentina, June 7th to 11th, 2015 (National Atomic Energy Commission of Argentina). P. 255–269.

25. Sarkisov G. N. Structural models of water // Uspekhi Fizicheskikh Nauk. 2006.Vol. 176. No. 8.

26. Fisher I.Z. Statistical theory of liquids. - M .: Fizmatgiz, 1961 .-- 280 p.

27. Eisenberg D., Kauzmann W. The Structure and Properties of Water. - Oxford: Clarendon Press. - 1969.

28. Sarkisov G. N. Approximate equations of the theory of liquids in the statistical thermodynamics of classical liquid systems // Uspekhi Fizicheskikh Nauk. 1999. V. 169, No. 6. P. 625–642.

29. Sarkisov G. N. Molecular distribution functions of stable, metastable and amorphous classical models // Uspekhi Fizicheskikh Nauk. 2002.Vol. 172, No. 6. P. 647–669.

30. Martynov G. A. Fundamental Theory of Liquids: Method of Distribution Function. - Bristol: A. Hilger. - 1992.

31. Stillinger F. H., Weber T. A. // Phys. Rev. A. 1982. V. 25. P. 978.

32. Nemethy G., Scheraga H. A. // J. Chem. Phys. 1962. V. 36. P. 3382.

33. Pauling L. in Hydrogen Bonding: Papers Presented. Symp., Ljubljana, Slovenia, 1957 (Ed. D Hadzi). - New York: Pergamon Press. - 1959.

34. Samoilov O. Ya. Structure of aqueous solutions of electrolytes and ion hydration. - M.: Publishing House of the Academy of Sciences of the USSR. - 1957.

35. Pople J. A., Proc R. // Soc. London Ser. A. 1951. V. 205.P. 163.

36. Rodnikova M.N. // Journal of Physical Chemistry. 1993. T. 67. C. 275.

37. Parfenyuk V. I. Some structural and thermodynamic aspects of solvation of individual ions. Thermodynamic characteristics of solvation of singly charged ions in water and methanol and their structural components // Journal of Structural Chemistry. 2001.Vol. 42, No. 6. S. 1133–1138.

38. Parfenyuk V.I., Paramonov Yu.A., Krestov G.A. // Dokl. USSR Academy of Sciences. 1990.Vol. 311, No. 1. P. 143–146.

39. Rabinovich VA. Thermodynamic activity of ions in solutions of electrolytes. - L .: Khimiya, 1985 .-- 176 p.

40. Parfenyuk V. I. Experimental methods of chemistry of solutions: densimetry, vis-

cometry, conductometry and other methods / V. K. Abrosimov, V. V. Korolev, V. N. Afanasyev, etc. - M .: Nauka , 1977. - S. 186–214.

41. Parfenyuk V. I. Diss. Doct. Chem. sciences. - Ivanovo: IHR RAS, 2000 .-- 189 p.

42. Parfenyuk V.I., Paramonov Yu.L., Chankina T.I. et al. // Dokl. USSR Academy of Sciences. 1988. T. 302, No. 3. P. 637–639.

43. Parfenyuk V.I., Chankina T.I. // Electrochemistry. 1994. Vol. 30, No. 6. P. 812–813.

44. Krestov G. A. Thermodynamics of ionic processes in solutions. - 2nd ed., Revised. - L .: Khimiya, 1984. - 272 p.

45. Krestov G. A., Abrosimov V. K. // Journal of structural chemistry. 1964. V. 4, No. 4. P. 510-515.

46. Abrosimov V.K. // Journal of Structural Chemistry. 1973.V. 14, No. 2. P. 211–217.

47. Entelis S. G., Tiger R. N. Kinetics of reactions in the liquid phase. - M.: Chemistry, 1973.- 416 p.

48. Jano I. // C. r. Acad sci. Paris 1965. V. 261, No. 7. P. 103-107.

49. Klopman G. K. // Chem. Phys. Lett. 1967. V. 1, No. 2. P. 200–203.

50. Tapia O., Goscinski O. // Mol. Phys. 1975. V. 29, No. 6. P. 1653–1661.

51. Tapia O., Sussman F., Poulain E. // J. Theor. Biol. 1978. V. 71. P. 49-58.

52. Tapia O., Poulain E., Sussman F. // Chem. Phys. Lett. 1977. V. 33, No. 1. P. 65–71.

53. Tapia O., Poulain E. // Intern. J. Quant. Chem. 1977. V. 11, No. 2. P. 473–479.

54. Tapia O. // Theor. chim. acta. 1978. V. 47, No. 2. P. 157–169.

55. Abronin I. A., Burshtein K. Ya., Zhidomirov G. M. // Journal of Structural Chemistry. 1980.Vol. 21, No. 2. P. 145–164.

56. Simkin B. Ya., Sheikhet I.I. // Physical chemistry: Modern. prob. - M.: Khimiya 1983. - P. 149-180.

57. 57. Daly L., Burton R. E. // J. Chem. Soc. Faraday Trans. Part II. 1970. V. 66, No. 7. P. 1281–1286.

58. Daly L., Burton R. E. // Ibid. 1971. V. 67, No. 7. P. 1219-1226.

59. Lischa H., Plesser Th., Schuster P. // Chem. Phys. Lett. 1970. V. 6, No. 2. P. 263–267.

60. Breitschwerdt K. G., Kistenmacher H. // Ibid. 1972. V. 14, No. 3. P. 283–291.

61. Diercksen G. H. F., Kraemers W. P. // Theor. chim. acta. 1972. V. 23, No. 3. P. 387–397.

62. Schuster P., Preuss H. // Chem. Phys. Lett. 1971. V. 11, No. 1. P. 35–39.

63. Kollman P. A., Kunz I. D. // J. Amer. Chem. Soc. 1972. V. 94, No. 26. P. 9236–9237.

64. Pullman A., Armbruster A.-M. // Chem. Phys. Lett. 1975. V. 36, No. 5. P. 558-561.

65. Cremaschi P., Simonetta M. // Theor. chim. acta. 1975. V. 37, No. 3. P. 341–347.

66. Port G. N., Pullman A. // Ibid. 1973. V. 31, No. 3. P. 231–240.

67. Kistenmacher H., Popkie H., Clementi E. // J. Chem. Phys. 1974. V. 61, No. 3. P. 799–810.

68. Popkie H., Clementi E. // Ibid. 1972. V. 57, No. 3. P. 1077-1084.

69. Kistenmacher H., Popkie H., Clementi E. // Ibid. 1973. V. 58, No. 5. P. 1969-1977.

70. Kistenmacher H., Popkie H., Clementi E. // Ibid. No. 12. P. 5627–5633.

71. Sord J. A., Probst M., Corongiu G. // J. Amer. Chem. Soc. 1987. V. 109, No. 6. P. 1702–1708.

72. Chandrasekhar J., Smith S. F., Jorgensen W. L. // Ibid. 1985. V. 107, No. 1. P. 154–163.

73. Madura J. D., Jorgensen W. J. // Ibid. 1986. V. 108, No. 10. P. 2517-2532.

74. Warchel A. // J. Phys. Chem. 1979. V. 83, No. 6. P. 1640-1650.

75. Burshtein K. Ya. // Journal of structural chemistry. 1987. V. 28, No. 2. P. 3–9.

76. Burshtein K. Ya. // Ibid. 1988. Vol. 29, No. 1. P. 179–182.

77. Andrade E. M., Dodd C. // Proc. Royal Soc. Sep. A. V. 187, No. 1946. P. 296–336.

78.  Sokolov P. T., Sosinsky S. L. // DAN USSR. 1937. T. 4, No. 127. S. 1037-1042.
79.  Dikarev V.M., Ostapenko A. L., Karasev G. G. // XI Intern. Conf. on conduction and Breakdown in dielectrics liquids. Zurich. 1993 (IEEE). P. 509-514.
80.  Shakhparonov MK. Methods of studying the thermal motion of molecules and the structure of liquids. - M .: Moscow State University, 1963 .-- 296 p.
81.  Richkov Yu.M., Lyon V.A. and others // Electronic processing of materials. 1994. No. 5. P. 34–37.
82.  Honda L., Sasada T. // Jap. J. Appl. Phys. 1977. V. 16, No. 10. P. 1775-1783.
83.  Bologa M. K., Gross F. P., Kozhuhar I. A. Electroconvection and heat transfer. - Chisinau: Shtiintsa, 1977 .-- 320 p.
84.  Ostapenko A. L. Influence of the electric field on the dynamic viscosity of liquid dielectrics // Journal of Technical Physics. 1998.V. 68, No. 1.
85.  Lutsky A.E. // ZhFH. 1978.Vol. 52, No. 4. P. 955–960.
86.  Boyko V.I., Kazaryan M.A., Shamanin I.V., Lomov I.V. // KSF FIAN. 2006. No. 7.
87.  Gusev A. L., Kazaryan M. A., Trutnev Yu. A. et al. // Alternative energy and ecology. 2007. No. 3.
88.  Baldanov M. M., Tanganov B. B., Mokhosoev M. V. // Dokl. USSR Academy of Sciences. 1988.V. 299. Issue. 4.
89.  Baldanov M. M., Tanganov B. B., Mokhosoev M. V. // Journal of Physical Chemistry. 1990.V. 64, No. 1.
90.  Baldanov M. M., Tanganov B. B. // Journal of Physical Chemistry. 1992. T. 66, No. 6.
91.   Fialkov Yu. Ya., Gorbachev V. Yu., Chumak VL. Conductometric determination of solvation numbers of alkali metal cations // Journal of Physical Chemistry. 1997. Vol. 71, No. 8. C. 1415-1419.
92.  Stokes G. G. // Trans. Camb. phil. Soc. 1845. V. 8. P. 287.
93.  Fialkov Yu. Ya. Solvent as a means of controlling the chemical process. - L .: Khimiya, 1990 .-- 240 p.
94.  Fialkov Yu. Ya., Zhitomirsky A.N. // Journal of Physical Chemistry. 1987.V. 64, No. 2. P. 390.
95.  Zwanzig R. J. // J. Chem. Phys. 1970. V. 52, No. 7. P. 3625.
96.  Habbard J., Onzager L. // J. Chem. Phys. 1977. V. 67, No. 11. P. 4850.
97.  Nightingale E. R. Jr. // J. Phys. Chem. 1959. V. 63. P. 1381.
98.  Gill D. S. // Electrochimica. Acta. 1979. V. 24, No. 6. P.701.
99.  Habbard J. // J. Chem. Phys. 1978. V. 68, No. 4. P. 1649.
100.  Stiles P. J., Habbard J. // Ibid. 1984. V. 84, No. 3. P. 431.
101.  Felderhof B. V. // Molec. Phys. 1983. V. 49, No. 2. P. 449.
102.  Novak E. // J. Chem. Phys. 1983. V. 79, No. 2. P. 976.
103.  Robinson R., Stokes R. Solutions of electrolytes: Per. from English - M .: Publishing house of foreign countries. lit., 1963. - 646 p.
104.  Marcus Y. // J. Sol. Chem. 1986. V. 15, No. 4. P. 291.
105.  Izmailov N. A. Electrochemistry of solutions. - M .: Chemistry, 1976 .-- 488 p.
106.  Baldanov M. M. // Izv. Universities. Series chem. and chem. technology. 1986.Vol. 29, no. 8.
107.  Baldanov M. M., Mokhosoev M. V. // DAN USSR. 1985.V. 284, no. 6.
108.  Zhigzhitova S. B. Influence of the properties of individual ions on the thermophysical characteristics of aqueous solutions of electrolytes in the framework of plasmahydrodynamic theory // Abstract of dissertation. - Ulan-Ude: Ed. SSSTU, 2007.
109.  Fedorov M. V., Kornishev A. A. Unravelling solvent response to neutral and charged solutes // Molecular Physics. 2007. V. 105 (1), No. 1.

110. Chiodo S., Chuev G. N., Erofeeva S. E., Fedorov M. V., Russo N., Sicilia E. // International Journal of Quantum Chemistry. 2007. V. 107 (2). P. 265.

111. Fedorov MV Development of the theory of solvation based on multiscale methods // Abstract of the dissertation. - Ivanovo, 2007.

112. Kazaryan M. A., Shamanin I. V., Lomov I. V. Physical models and applications of the process of solvation of salt ions in polar dielectrics // Alternative Energy and Ecology. 2007. No. 11.

113. Nevolin C. K. Physical foundations of tunnel probe nanotechnology // Electronic industry. 1993. No. 10. P. 8-15.

114. Zhakin A. I. Ionic conductivity and complexation in liquid dielectrics // Uspekhi Fizicheskikh Nauk. 2003.Vol. 173. No. 1.

115. Stishkov Yu. K., Ostapenko A. A. Electrohydrodynamic flows in liquid dielectrics. - L .: Publishing house of Leningrad State University, 1989.

116. Rychkov Yu.M., Stishkov Yu.K. // Colloidal Journal. - 1978. No. 6. P. 1204.

117. Izmailov N. A. Electrochemistry of solutions. - M.: Chemistry, 1966.

118. Bjerrum N. K. Dan. Vidensk. Selsk. Mat.-Fys. Medd. 1926. V. 7. No. 9.

119. Qian Xue-Sen Physical mechanics. Translation from Chinese, ed. R. G. Barantseva. - M.: Mir, 1965.

120. Atrazhev V. M., Timoshkin I. V., in Proc. of 1996 IEEE 12th Intern. Conf. on Conduction and Breakdown in Dielectric Liquids, Rome, Italy, July 15–19, 1996 (New York: IEEE, 1996) P. 41.

121. Khrapak A. G., Volykhin K. F., in Proc. 12th Intern. Conf. on Conduction and Breakdown in Dielectric Liquids (Rome, Italy, 1996). - P. 29.

122. Terenin AN // Successes in physical sciences. 1937. T. XVII, no. 1.

123. Vlaev L.T., Nikolova M.M., Gospodinov G.G. Electric transport properties of ions in aqueous solutions of H2SeO4 and Na2SeO4 // Journal of Structural Chemistry. 2005. T. 46, No. 4. P. 655–662.

124. Samoilov O. Ya. Structure of aqueous solutions of electrolytes and ion hydration. - M.: Publishing House of the Academy of Sciences of the USSR, 1957.

125. Robinson R., Stokes R. Solutions of electrolytes. - M .: Publishing house of foreign countries. lit., 1963.

126. Griliches M.S., Filanovsky B.K. Contact conductometry. - L .: Chemistry, 1980.

127. Gill D. S. // Electrochimica Acta. 1977. V. 22, No. 4. R. 491–492.

128. Gill D. S. // Ibid. 1979. V. 24, No. 6. R. 701–703.

129. Fialkov Yu. Ya., Gorbachev V. Yu., Chumak V.L. // Journal of Physical Chemistry. 1997.V. 71, No. 8. P. 1415-1419.

130. Nightingale E. R. // J. Phys. Chem. 1959. V. 63. P. 1381–1387

131. Samoilov O. Ya., Uedira H., Yastremsky P.S. // Journal of Structural Chemistry. 1978. T. 19. No. 5. S. 814–817.

132. Vlaev L.T., Genieva S.D. // Journal of Physical Chemistry. 2003.Vol. 77. No. 12. P. 2178–2183.

133. Vlaev L. T., Genieva S. D., Tavlieva M. P. The concentration dependence of the activation energy of the specific conductivity of aqueous solutions of sodium selenite and potassium tellurite // Journal of Structural Chemistry. 2003.V. 44. No. 6. P. 1078-1084.

134. Engel G., Hertz H. G. // Ber. Bunsengesel. Phys. Chem. 1968. V. 72, No. 7. P. 808–834.

135. Valyashko V. M., Ivanov A. A. // Journal of Inorganic Chemistry. 1979.Vol. 24, No. 10. P. 2752–2759.

136. Vasilieva L. F., Gitis E. B., Shmorgun V. I. // Journal of Physical Chemistry. 1976.V.

49, No. 11. P. 2539–2541.

137.  Samoilov O. Ya. Structure of aqueous solutions and ion hydration. - M.: Publishing House of the Academy of Sciences of the USSR, 1957.

138.  Pamfilov V.A., Kuzub V.S., Kuzub L.G. // Ukr. Chem. journal 1960.V. 26, No. 2. S. 174–181.

139.  Pamfilov A.V., Dolgaya O.M. // Journal of Physical Chemistry. 1963. T. 37, No. 8. P. 1800–1804.

140.  Ivanov A.A. // Izv. UNIV.. Chemistry and Chem. technologist. 1989.V. 32.S. 3–16.

141.  Krestov G. A., Abrosimov V. K. // Journal of structural chemistry. 1967. V. 8, No. 5. P. 822–826.

142.  Smolyakov B. S., Veselova G. A. // Electrochemistry. 1975.Vol. 11, No. 5. P. 700–703.

143.  Yergin Yu. V., Kostrova L.I. // Journal of Structural Chemistry. 1971.V. 12, No. 4. P. 576-579.

144.  Rodnikova M.N., Nosova T.A., Markova V.G., Dudnikova K.T. // Dokl. RAS. 1992. V. 327, No. 1. S. 96–99.

145.  Zasypkin S. A., Rodnikova M. N., Malenkov G. G. // Journal of Structural Chemistry. 1993. V. 34, No. 2. P. 96–104.

146.  Rodnikova M.N. // Journal of Physical Chemistry. 1993. V. 67, No. 2. P. 275–280.

147.  Stishkov Yu. K., Steblenko A.V. // Journal of Technical Physics. 1997.V. 67. No. 10.

148.  Kazaryan M. A., Shamanin I. V., Lomov I. V. et al. Sizes of solvated ions (clusters) in salt solutions // Brief Communications on Physics, Lebedev Physical Institute. 2007. No. 8. P. 35–43.

149.  Shamanin I. V., Kazaryan M. A. Clusters Formation in Salts Solution in Polar Dielectric Liquids and Electrically-induced Separation of Solvated Ions // British Journal of Applied Science and Technology. 2014. Vol. 4, No. 18. P. 2538–2550.

150.  Landau L. D., Lifshits E. M. Electrodynamics of continuous media. - M: Fizmatgiz, 1973.- 454 p.

151.  Frank-Kamenetsky D. A. // Lectures on plasma physics. - M .: Atomizdat, 1968.

152.  Landau L. D., Lifshits E. M. Theoretical physics: Vol. 1. Mechanics. - M .: Nauka, 1988 .-- 216 p.

153.  Frenkel Ya. I. Kinetic theory of liquids. - L .: Nauka, 1975 .-- 592 p.

154.  Eisenberg D., Kautzman V. Structure and properties of water. - L .: Gidrometeoizdat, 1975.

155.  Delone N. B. Interaction of laser radiation with matter. - M: Nauka, 1989 .-- 373 p.

156.  156. Delone N. B., Kraynov V. P. Nonlinear ionization of atoms by laser radiation - M: Fizmatlit, 2001. - 421 p.

157.  Krainov V.P. Orientation and focusing of molecules by the laser radiation field // Soros Educational Journal, 2000, No. 4. P. 90–95.

158.  Feynman R., Leighton R., Sands M. Feynman lectures on physics. T.5. Electricity and magnetism. / Translation from English. G.I. Kopylova, Yu.A. Simonova. Ed. J. A. Smorodinsky. - M .: Mir, 1977 .-- 302 p.

159.  Boyko V. I., Vlasov V. A., Zherin I. I. et al. Thorium in the nuclear fuel cycle. - M.: Ore and Metals, 2006 .-- 358 p.

160.  Stepin B. D., Gorstein I. G., Blum G. Z. et al. Methods for the production of highly pure inorganic substances. Publishing House "Chemistry", Leningrad Branch, 1969. - 480 p.

161.  Belashchenko D.K. Electric transport in liquid metals. // Advances in chemistry. 1965.V. 34, no. 3. P. 530–564.

162.  Shemlya M. Separation of isotopes .: Per. with french / M. Shemlya, J. Perrier. - M

.Atomizdat, 1980 . 169 p.

163. Wilson, J. R. Demineralization by electrodialysis. // Trans. from English under the editorship of B.N. Laskorin and F.V. Rausen. - M .: Gosatomizdat, 1963 .-- 351 p.

164. Dukhin S. S. Electro-surface phenomena and electro-filtering / S. S. Dukhin, V. R. Estrella-Llopis, E. K. Zholkovsky; Academy of Sciences of the Ukrainian SSR; Institute of Colloid Chemistry and Water Chemistry. - Kiev .: Naukova Dumka, 1985 .-- 287 p.

165. Electrokinetic properties of capillary systems: a monographic collection of experimental studies performed under the guidance of corresponding students of the USSR Academy of Sciences I. I. Zhukov / USSR Academy of Sciences, Department of Chemical Sciences; under the editorship of P. A. Rebinder. - M .; L .: Publishing House of the Academy of Sciences of the USSR, 1956. - 352 p.

166. Andryushchenko F.K. Theoretical Electrochemistry: Textbook / F.K. Andryushchenko, V.V. Orekhova. - Kiev .: Vishcha school, 1979. - 167 p.

167. Kargin V. A., Lastovsky R. P., Matveeva T. A. et al. Purification of titanium dioxide and metatitanic acid by high-voltage electrodialysis. // LC. 1961.Vol. 6, no. 5. S. 1017–1019.

168. Karchevsky A. I., Martsinkyan V. L., Popov I. A. et al. Separation of xenon isotopes in a high-frequency gas discharge. // Plasma physics. 1977. T. 3, No. 2. P. 409-417.

169. Gorbunova EF, Ezubchenko A. N., Karchevsky A. I. et al. Separation of xenon isotopes in a stationary high-frequency discharge with a traveling electromagnetic wave. // Letters to the ZhTF. 1977. T. 3, No. 4. P. 154–157.

170. Babichev A. P., Gorbunova E. F., Yezubchenko A. N. et al. Separation of isotopes of inert gases in a stationary high-frequency discharge with a traveling magnetic field. // Letters to the ZhTF. 1979. No. 9. P. 1872–1878.

171. Gorbunova E. F., Karchevsky A. I., Muromkin Yu. A. Isotope separation in the positive column of a gas discharge. // Plasma physics. 1986.Vol. 12, no. 9, p. 1087.

172. Gorbunova E. F., Ezubchenko A. N., Karchevsky A. I. et al. Separation of isotopes of inert gases in a stationary high-frequency discharge in a traveling magnetic field. // ZhTF. 1979.V. 49, no. 9.P. 1872–1878.

173. Laranjeira M., Kistemaker J. Experimental and theoretical thermal diffusion factors in gaseous mixtures. // Physica. 1960. V. 26, No. 6. P. 431.

# Index

## A

approximation
  Debye–Hückel approximation  viii,  117
  supermolecular approximation  87
aquacomplex  50,  51,  52,  53,  54,  55,  56,  57,  67,  68,  69,  70,  71,  72,  73,  74
    ,  78

## B

Brownian collisions  48,  58,  118

## C

Clementi potentials  92
cluster  9,  43,  46,  58,  66,  67,  68,  70,  74,  75,  76,  77,  79,  82,  101,  110,  113,
    118,  119,  120,  124,  125,  126
clusters  1,  48,  49
coefficient
  surface tension coefficient  74,  75
concept  78,  81,  86,  104,  108,  120
  Semenchenko –Bjerrum concept  120

## D

dielectric permittivity  3,  166,  167,  168,  169
drift
  electroinduced drift  43,  47,  50,  126,  129,  144,  179
  oriented drift  25,  26
  selective drift  1,  2,  6,  18,  20,  21,  33,  35,  36,  50
  translational drift  7

## E

effect
  electrohydrodynamic  100
  electroviscous effect  100
  optical Kerr effect  171,  172
  orientation Kerr effect  170
electric field vector  53,  112
electrodialysis  183